Fire Command
Instructor's Guide

to

Fire Command Second Edition

by

Alan V. Brunacini and Nick Brunacini

NFPA®

National Fire Protection Association
Quincy, Massachusetts

Fire Command Instructor's Guide
for Fire Command Second Edition

Written by Alan V. Brunacini and Nick Brunacini

Designed and Edited by:
Uptown Graphics and Design
9143 W. Lone Cactus Drive
Peoria, AZ 85382
(623) 537-0419

NFPA No. FCINSTRGD
ISBN 087765-555-3
Library of Congress Control Number 2003107513
Copyright 2003 Alan V. Brunacini and Nick Brunacini

INTRODUCTION

The FIRE COMMAND Incident Management System (IMS) is used by fire departments and other emergency response agencies, on an international scale, for more than two decades. This second edition of FIRE COMMAND has been designed to be the foundation of a complete curriculum package. This package includes:

- the FIRE COMMAND Textbook
- the FIRE COMMAND Workbook
- the FIRE COMMAND Instructor's Guide
- a CD containing three different PowerPoint programs:

  - FIRE COMMAND PowerPoint presentation. This presentation outlines the personal traits required to be an effective IC and covers each of the bullet points in the IC Checklist, found at the end of each chapter (one through eight) in the textbook. These checklists cover the core competencies for the material found in each chapter (one through eight). This is a really fancy way of saying, "the system works if the IC does the bullet points when s/he is in command of an incident ."

  - Simulation PowerPoint presentation. This program contains simulated fires in eighty-two different buildings. Many of the simulations are four set series--the same view of the structure with four "levels" of conditions: nothing showing, offensive conditions, marginal conditions, defensive conditions. Some of the simulations include views from multiple sides of the occupancy along with interior shots. All of the simulations include building layout (similar to what the IC would draw on a tactical work sheet) and an overview page that gives the buildings an address, square footage and basic construction features.

  - Pyroville. This program is a drawing of a typical community (hence "Pyroville"). Pyroville covers approximately six blocks and includes thirty-three slides of twelve different buildings. This presentation is designed to be projected onto a white board. Based on the lesson or the point to be made, the instructor has the option of coloring the fire and smoke in the appropriate location. It is intended to be a simulate as you go program.

- The set of nine FIRE COMMAND videos. This video package includes an introductory video that provides a comprehensive overview of the IC's role in the Incident Management System and one video for each of the eight functions of command. Each of the eight standard functions of command videos goes into the specifics of how the IC performs that particular function.
- Instructor's Video. This video describes how the different pieces of the curriculum package fit together and provides teaching options for the instructor.

The FIRE COMMAND Incident Management System is highly customizable to local conditions. This system is used by large metropolitan fire departments, small volunteer fire departments, departments of correction, industrial fire brigades, law enforcement agencies, hazardous materials response teams, technical rescue teams, and EMS providers. The key to managing effective and safe emergency service delivery is by consistently using an incident management system that fits your local response. The way we accomplish this on a local level is to determine what our standard response (i.e., personnel and resources) will be and design our IMS to manage that response. The Up-Front Stuff chapter of FIRE COMMAND describes all the different elements that a local jurisdiction should consider when designing, building, implementing and maintaining their own local system.

The authors of the FIRE COMMAND package are teachers (and students) of incident management. They have been actively (and professionally) involved throughout their careers in the care and feeding of the incident management system used in a very large and busy metropolitan area. The FIRE COMMAND system is used by the twenty-two fire departments in their valley, and all the other incident responders, who protect a 1200 square mile area that four million people call home.

This teaching package was designed around the needs of the instructor. This curriculum is used in Universities, Community Colleges and fire department training programs around the world. This edition makes life easier for the instructor--everything you need to teach the class is included in this package. The instructor's video provides a good, informal overview of how the curriculum package works.

We would like to take this opportunity to thank you for using the FIRE COMMAND system. We can be contacted at firecommand@cox.net with any questions or comments. Be safe.

Table of Contents

Fire Command Instructor's Guide

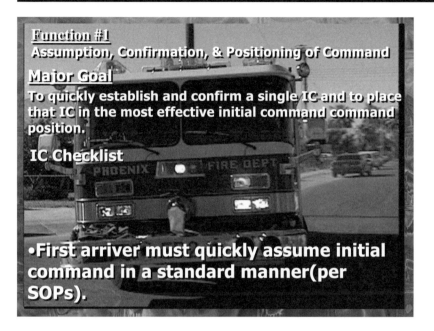

Function #1
Assumption, Confirmation, & Positioning of Command
Major Goal
To quickly establish and confirm a single IC and to place that IC in the most effective initial command command position.
IC Checklist

•**First arriver must quickly assume initial command in a standard manner(per SOPs).**

■ First arriver must quickly assume initial command in a standard manner (per SOPs)

The basic IMS must be well understood by all the participants. If the operational and command system is a big mystery to the workers that use it, it simply won't work--that's why we say assume command in a standard manner. Standard system stuff is written down, practiced, used and revised as needed. This is what causes it to be a standard regular piece of incident operations.

If the system doesn't kick in until some later-arriving "command type" (i.e., the chief) shows up, everyone assigned to the incident before s/he arrives is basically operating outside the system. Operating this way is frustrating for the person who eventually assumes command and it is very dangerous for the workers that operate at an uncommanded/unmanaged event.

Some people are uncomfortable with a non-officer assuming command (in cases where they are the first arriving member). It makes perfect sense for the first arriver (regardless of rank) to assume command because at that initial point, no one else from the fire department is on the scene. When someone else arrives to the scene and they have a better ability to command the event (through experience, rank, where they will be physically positioned--i.e., command post, or they have gypsy blood) then they can transfer command. That is the way the system is designed.

Simulation Questions

1. **Who will be the initial IC?**
   The first arriving responder
2. **When will they become command?**
   At the very beginning of the event
3. **Where will the initial IC be physically located?**
   Depending on rank and role in one of three modes:
   * *Nothing-Showing mode*
   * *Fast-Action mode*
   * *Command mode*
4. **How will they take command?**
   By broadcasting a standard initial radio report including... the assumption of command
5. **Describe the organizational effect that assuming command has on the incident operation.**
   Responders who arrive after command is established:
   * *Follow staging SOPs*
   * *Staged units receive and acknowledge orders from the IC*
   * *Go to work in the IC's order/work under IC's command*

Fire Command Instructor's Guide

Simulation Questions

1. **Do incident operations begin with an in place IC?**
   - *...ragged starts are prevented, simply by having our first responder always capture command at the very start...*

2. **Does the IC "out-rank" all the incident participants?**
   - *...that person (who establishes "command") outrank everyone else in the system during the time they are serving as the IC.*

3. **Is the IC "empowered" to make strategic level decisions and request (call for more resources, write off lost property, etc.)?**
   - *The command system empowers and directs the first arriver...to begin performing the functions of command*

4. **Is the initial (and subsequent) IC responsible for performing the 8 functions of command?**
   - *The current IC is always responsible for performing the standard command functions.*

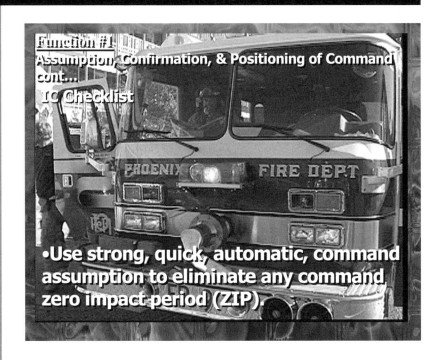

- Use strong, quick, automatic, command assumption to eliminate any zero-impact period (ZIP).

  Assuming command causes the initial arriving member (the IC) to size up the incident, determine the strategy, and formulate an incident action plan (IAP). All of this is executed and shared with all the incident participants when the IC transmits the initial radio report. This puts all the incident players on the same page. Everyone knows what the problem is and what action is being taken to solve it. Quick, effective and coordinated initial action is the result of a well-managed beginning, eliminating wasted effort in the critical first few minutes of the event.

  The absence of an effective IC is the most common reason for ragged incident beginnings. Effective (and coordinated) action is the result of beginning (and ongoing) incident operations with an in place and in charge IC.

Fire Command Instructor's Guide

Assume, Confirm and Position Command
Bullet point #1-3

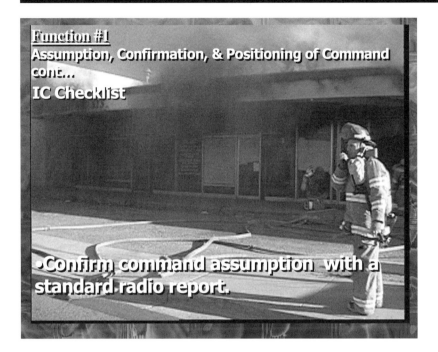

Function #1
Assumption, Confirmation, & Positioning of Command cont...
IC Checklist

•Confirm command assumption with a standard radio report.

■ Confirm command assumption with a standard initial radio report.

This is very important. When the first unit arrives to the scene, gives an initial radio, but doesn't say they are assuming command it screws the event up from the start. The most pressing problem for the responders is no longer the incident problem, it's who is in command of the event (managing and coordinating all the action that must take place). There should never be any doubt who the IC is (or that there is an IC). We use the system to manage effective action, the IC is the person responsible for managing the system. No IC equals no effective action. Whoever gives the initial radio report must include that they "will be command" as part of that report.

Another powerful benefit that that the onscene report provides to the IC is it wraps the first five functions of command together. This is particularly important when the Company Officer IC is operating in the fast attack mode and time is at a premium.

Simulation Questions

1. What is your onscene report for the structure fire pictured above?
2. Does the onscene report address:
   - *Chosen strategy*
   - *Identify incident problem*
   - *Initial IAP and action taken*
   - *Unit assuming command*
3. Describe the effect that the initial onscene report has on the rest of the incident responders (use the functions of command).

Fire Command Instructor's Guide

Simulation Questions

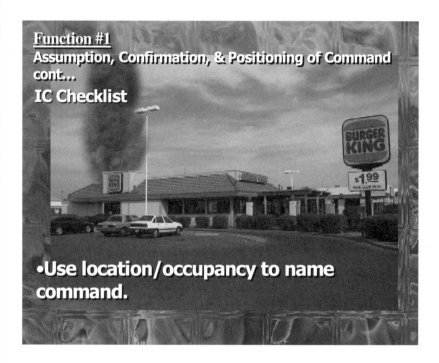

Function #1
Assumption, Confirmation, & Positioning of Command cont...
IC Checklist

•Use location/occupancy to name command.

1. The fire occupancy is located on Main Street. What will you name command?
2. Does the IC lose the radio ID of their unit and use the radio designation "_____ Command"?
3. Do subsequent arriving ICsfollow this same model?

■ Use location/occupancy to name command.

When we name command it identifies the IC of that particular incident. Everyone assigned to the incident then knows they are working for "Main Street Command" (as an example). When Main Street Command radios the Communications Center, to ask for another alarm, the Communications Center will know to send more resources to the Main Street incident and not the Elm Street incident that is being worked on the other side of town.

Naming command also has a strong organizational effect. Over time, it moves the system from being driven by specific individuals to one that is system based. Example:

Incident operations don't run smoothly until Chief Smith shows up to the scene and really assumes command. In this organization, the command system actually begins when Chief Smith arrives to the scene and announces over the tactical, "Car 3 (or whatever Chief Smith calls himself) is on the scene and will be Command". In this case Command (Chief Smith) will usually maintain his normal ID for radio communications--"Car 3 to dispatch…".

Fire Command Instructor's Guide

Assume, Confirm and Position Command
Bullet point #1-4

Simulation Questions

Organizations that have made training investments in developing their local IMS and in training their responders move away from this type of individualized command capability. When the initial arriver assumes command by announcing over tactical channel, "…Engine 1 will be Main Street Command," the focus shifts from the individual (the person who is command) to the fact that an IC has been established and that person is now responsible for doing the functions of command.

Fire Command Instructor's Guide

Assume, Confirm and Position Command
Bullet point #1-5

Simulation Questions

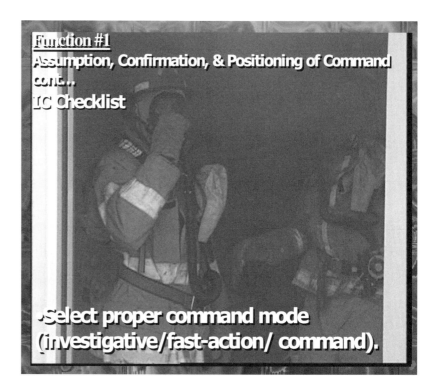

Function #1
Assumption, Confirmation, & Positioning of Command cont...
IC Checklist

Select proper command mode (investigative/fast-action/ command).

1. What command mode will the initial IC operate in?
2. Is the chosen command mode rank specific--can chief officers operate in all three command modes, how about company officers?
3. List the advantages and disadvantages of the different command modes for company officers.
4. List the advantages and disadvantages of the different command modes for chief officers.

■ Select the proper command mode (investigative/fast-action/ command).

The proper command mode will be based on a variety of factors. They include the chosen strategy, the tactical capabilities of the initial arriving unit, and the rank and role the initial IC plays as a member of the initial arriving company/unit. An Officer IC of an initial-arriving engine company that chooses the offensive strategy will most likely operate in the fast attack mode. If a Battalion Chief is the initial-arriving unit to the same structure fire, they will operate in the command mode (chief officers, who don't arrive on a fire fighting unit, i.e., engine, ladder, should always operate in the command mode when they are the IC).

Fire Command Instructor's Guide

Function #1
Assumption, Confirmation, & Positioning of Command cont...
IC Checklist

•Correctly position command to match & support the current command mode.

Simulation Questions

■ Correctly position command to match and support the current command mode.

This means if the initial IC chooses the fast attack mode they should be inside the hazard zone leading the problem solving attack--not outside with their turnout coat unbuttoned, standing in the front yard. If a battalion chief is the IC operating in the command mode they are outside running the incident from inside a stationary and remote command post--they are not inside the hazard zone wearing their Class A uniform (or even their full protective gear).

1. What is the information gathering advantage of the initial IC evaluating conditions from the exterior of the fire occupancy and then on the interior of the building (in the offensive strategy)?

2. How does the IC receive information from all the different positions around the incident scene when operating from a fixed and static location, when operating in the command mode?

Assume, Confirm and Position Command
Bullet point #1-7

Simulation Questions

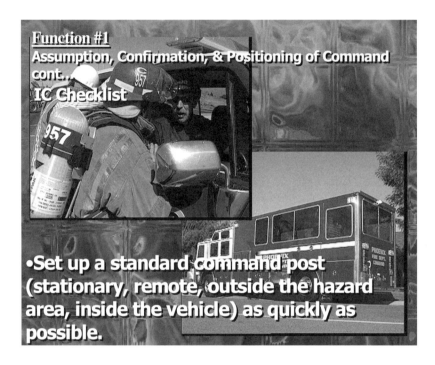

Function #1
Assumption, Confirmation, & Positioning of Command cont...
IC Checklist

•Set up a standard command post (stationary, remote, outside the hazard area, inside the vehicle) as quickly as possible.

1. How many resources can an IC that is operating in the fast attack mode assign to the incident.
2. How well can they manage the position and function of those resources?
3. At what point does the fast attacking IC become overwhelmed and ineffective?
4. At what point in the incident operation is it mandatory to have an IC that will be operating in the command mode?

■ Set up a standard command post (stationary, remote, outside hazard area, inside the vehicle) as quickly as possible.

The most effective position to command from is inside a command post, not a burning building. The system has got to be quick and automatic, this requires (in many cases) the initial IC (a company officer) to mix action with being in command. This should end as quickly as possible in one of two ways--the initial attack solves the problem--eliminating the hazard or command is transferred to an IC that will be operating in the command mode, inside a command post.

Fire Command Instructor's Guide

Function #1
Assumption, Confirmation, & Positioning of Command cont...
IC Checklist

•Begin to package command for ongoing operation and escalation:
- **strong standard command**
- **sectors**
- **SOPs**
- **clear communications**
- **standard strategy/action planning**

Simulation Questions

1. List the set of problems that can occur if the initial IC doesn't "package" command and the incident operation continues to escalate.
2. List the set of items the IC #2 will have to address/correct if IC #1 didn't "package" command:
   - *Begin to "package" command for on-going operation and escalation:*
     - *-Strong standard command*
     - *- Sectors*
     - *- SOPs*
     - *- Clear communications*
     - *- Standard strategy/ action planning*

■ The preceding seven bullet points put an effective IC in charge of the incident. Once in place the IC can now use the "standard" pieces of the incident management system to control incident operations. This boils down to the IC always being in a position where s/he can control where the workers are along with matching their actions to the incident conditions. The IC maintains this capability by decentralizing the incident scene by assigning sector officer responsibilities. When the IC orders an engine company to "lay a supply line to the north side and pull a attack line for fire control" s/he knows what that evolution looks like because of the SOP everyone uses to lay lines and pull attack line (it makes it a standard event). This becomes the core for the orders the IC communicates to companies/sectors. Identifying the strategy and action plan describes to everyone where the fight will take place and the tactics we will use to pull it off.

Packaging command in this fashion, from the onset of incident operations, provides a seamless transfer when command is transferred from the initial IC to a later arriving command officer. This allows IC #2 to continue the incident operation without having to fix incident management related problems.

Fire Command Instructor's Guide

Simulation Questions

1. How many units can a fast attacking IC assign and manage at this fire?
2. Describe the transfer of command process that takes place between the initial IC and IC #2.
3. The most effective way to transfer command is by using the tactical worksheet to exchange critical information. This is not possible when transferring command from a fast attacking IC. Describe the process used to transfer command when reviewing the tactical worksheet isn't possible.

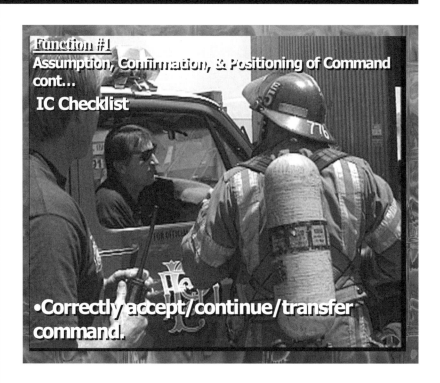

Function #1
Assumption, Confirmation, & Positioning of Command
cont...
IC Checklist

•Correctly accept/continue/transfer command.

■ Correctly accept/continue/transfer command.

This ensures that an effective IC will begin the event and the level of needed command will escalate based on the size and hazard of the incident. This becomes the core of the system, always matching the right amount of command to the incident. It eliminates any command gaps (ZIP) from occurring. In most cases incident operations will begin with a company officer assuming command. When the chief (or whomever your local system uses in that role) arrives to the scene and transfers and upgrades command. If the incident continues to escalate the IC (and the system) add whatever command support (Command Teams, Section Chiefs, etc.) is required to effectively and safely manage the incident.

Fire Command Instructor's Guide

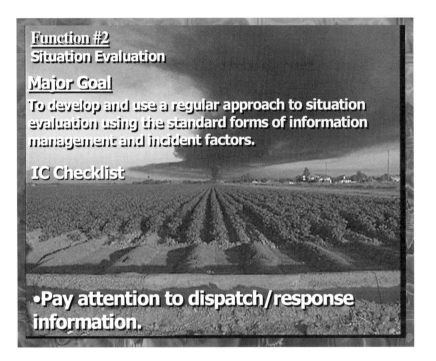

Function #2
Situation Evaluation

Major Goal

To develop and use a regular approach to situation evaluation using the standard forms of information management and incident factors.

IC Checklist

•Pay attention to dispatch/response information.

Simulation Questions

1. What is critical dispatch information --what do you want to know on the dispatch?
2. When do you start formulating your incident action?
3. Will you request more resources based on the incident address and apparent incident conditions (i.e., smoke/flame showing) during your response?

■ Pay attention to dispatch information.

Dispatch information includes the location of the incident (address), what the incident problem is, what other units are assigned to the incident and the radio channel the incident will be worked on. If the incident is a fire call everyone can get turned out prior to leaving (this is one less thing to have to do while enroute--get dressed prior to getting on the rig makes it easier to wear your seat belt). The driver should be certain the entire crew is seated and belted and that the bay door is all the way up and know both the destination and route before the apparatus moves. The first thing the officer should check after getting on the rig is that the radio is on the correct channel.

The address provides responding crews a general idea of the type of structure and occupancy they are responding to (commercial area, multi-story apartments, etc.). It also lays out the arrival order of all of the responders. If Engine 1 is the first due unit on a multiple unit response the members of the later arriving companies are listening attentively to the radio for Engine 1 to clear the communications center to give the initial radio report (especially if there is smoke on the horizon).

Simulation Questions

1. Discuss the information gathering pros and cons for the fast attack IC – what are the advantages and disadvantages doing a size up on the exterior of the building then moving to the interior (for offensive fires)?
2. Are the basic critical incident factors the same from fire to fire?
3. Is it possible that at certain incidents one or more of the critical factors may be so significant that it changes the order in which we address/complete the tactical priorities?

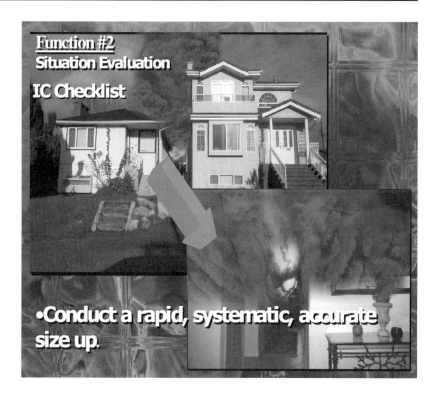

Function #2
Situation Evaluation
IC Checklist

•Conduct a rapid, systematic, accurate size up.

■ Conduct rapid, systematic, accurate size-up.

The initial IC must do the first five functions of command within a very short time frame. None of this becomes any easier when you consider that in most cases the initial IC is the Officer of an engine company--the only person in the incident scene organization that operates on the task, tactical and strategic levels all at the same time. The size-up of the incident critical factors is the basis for choosing the proper strategy and formulating the incident action plan. The initial IC only has time to consider a handful of critical incident factors--things like fire size, extent and location, the occupancy type, and the construction characteristics of the structure as they relate to the fire.

This initial size-up should be based on those critical factors, the risk management plan (for firefighter safety) and completing the tactical priorities. Using this type of standard size-up system provides the IC a powerful and effective tool to match actions to conditions and serves as the basis for setting priorities and making and managing assignments based on those priorities.

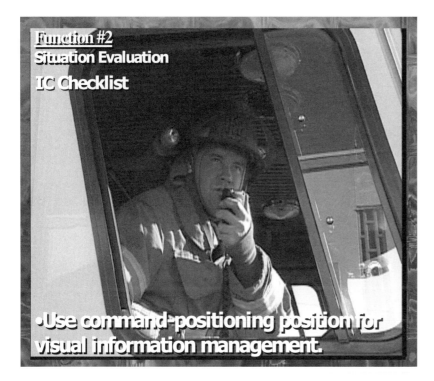

Function #2
Situation Evaluation
IC Checklist

•Use command-positioning position for visual information management.

Simulation Questions

1. What is the best command position (or mode) for ongoing visual information management?
2. Is the interior of a burning building a good position to size up incident factors for the entire incident site?
3. How does the IC know/find out if the fire has extended into the exposures on either side of the fire occupancy?

■ Use command-position for visual information management.

When the initial IC chooses the offensive strategy and the fast attack mode s/he makes their initial size-up from and exterior position. They look at the effect the incident problem is having on the outside of the hazard zone. The IC then moves into an interior (fast attack) position and begins collecting information of how the incident problem is effecting the inside of the structure. Heavy smoke, high heat (or light smoke and no heat) then get processed back into the decision making process. The goal is to collect all the needed information that is required to effectively solve the problem while keeping firefighters safe.

When the IC is operating in the command position they will normally have a good view of the incident scene. As the IC assigns units to the different operational positions around the incident scene he/she will get back size up information in the form of progress/recon reports from these different positions. As these reports come back to the IC, s/he combines them with what s/he is actually seeing. This combination of visual and recon information paints a picture (size up) of the incident operation and associated hazards.

Fire Command Instructor's Guide

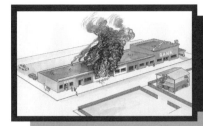

Simulation Questions

1. Does the initial IC need much in the way of preplan info to initiate incident operations for this fire?
2. What information would you include in a preplan of this building?
3. What incident players are in the best positions to use preplans?

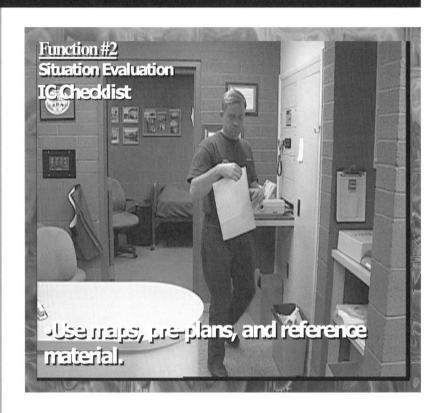

Function #2
Situation Evaluation
IC Checklist

•Use maps, pre-plans, and reference material.

■ Use maps, pre-plans, and reference material.

This recorded pre-loaded, pre-plan information is only as useful as it is accessible. If it takes ten minutes to find the right book and dig the preplan out, it will not be used by the initial IC (fast attacking company officer). If the reference material is stored in a way that it can be quickly accessed it is much more likely that it will be used. This material should only include the information needed to make effective firefighting decisions. A set of blue-prints (as an example) for the structure is overkill and will not be used by anyone wearing full protective gear (or anyone operating in the command post for that matter). Pre-plans and other reference material needs to be short, spccific, to the point and easy to get to. The computer age will (and in some places has) revolutionized information management.

Situation Evaluation
Bullet point #2-5

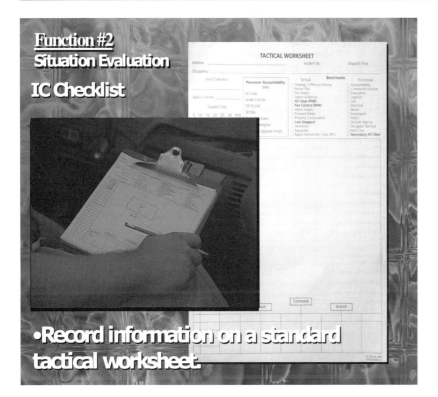

Function #2
Situation Evaluation

IC Checklist

•Record information on a standard tactical worksheet.

Simulation Questions

1. When does the IC begin to fill out the tactical worksheet?
2. Is it SOP for the initial arriving IC (the Company Officer of the initial arriving unit) to begin a tactical worksheet?
3. How does the fast attacking IC keep track of the resource he/she assigns?
4. Does the tactical worksheet provide the IC with a tool to better manage the incident?
5. Can "Command" become screwed up if the IC/Command Team doesn't maintain an accurate and update tactical worksheet?

■ Record information on a standard tactical worksheet.

Tactical worksheets should be filled out and updated in the command post. A fast attacking IC will not use a tactical worksheet--paper and pencil are not designed to be used in the hazard zone. The fast attacking IC shouldn't be expected to manage more units than s/he can keep track of off the top of their head. In most cases this number will be between 3-4.

A simple line drawing on the TWS of the building, noting the assigned positions and locations of operating companies allows an IC, that is operating in the command mode, to track the resources required beyond a first alarm (4 engines, 2 ladders, and support units). Try remembering where you assigned 6 engines, 2 ladders, and 4 pieces of support equipment over a 15-minute time frame, if you didn't write it down somewhere--good luck.

The tactical worksheet also provides a checklist (memory joggers) for the tactical priorities and key tactical benchmarks. A properly filled out worksheet not only shows the IC where his/her troops are but also any uncovered areas that need to be addressed.

Situation Evaluation
Bullet point #2-6

Simulation Questions

Function #2
Situation Evaluation
IC Checklist

•Use companies and sectors as information, reporting, and recon agents.

1. List all the positions at this fire that the IC needs recon information from--
2. How does the IC get this info?
3. How do Sector Officers package that info and get it back to the IC (or who ever needs it)?
4. How does the IC manage conflicting information--what he/she is seeing doesn't match up to the report s/he is receiving?

■ Use companies and sectors as information, reporting, and recon agents.

As the IC assigns companies and sectors to key operating positions they must report back about the conditions in their assigned areas. The IC uses this information to build a strategic picture of what is happening around the entire incident site. This "picture" is what the IC uses to keep the strategy and attack plan current.

The basis of this two-way commo is an agreement among all the team members on what the standard critical factors are, a common language on how we describe those factors to each other, and a mutual understanding about what is the standard (effective/safe) response to those factors. Both ends of the process (sender/receiver) must be able to identify and describe the level of intensity of that factor and be connected to how that intensity is changing--either getting better (safer) or worse (more hazardous). The longer the team practices this info exchange, the more effective the system becomes.

Situation Evaluation
Bullet point #2-7

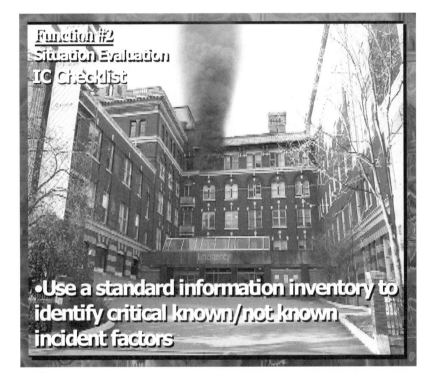

Function #2
Situation Evaluation
IC Checklist

•Use a standard information inventory to identify critical known/not known incident factors

Simulation Questions

1. Answer the questions (1-8) for the fire pictured above.

■ Use a standard information inventory to identify critical known/not known incident factors.

This information inventory includes the critical factors that relate to the IC's ability to complete the tactical priorities and the factors that effect firefighter safety. A basic list for structure fires looks like this:

1. What type of fuel, and how much, is on fire (class A, B, C, D)?
2. How much is left to burn?
3. What type of occupancy (house, commercial, fireworks factory, etc)?
4. What are the conditions in the occupied/living/working spaces?
5. What are the conditions in the concealed spaces (attic, basement, etc.?
6. Is the structure stable and how is the fire effecting it?
7. If you can get the troops in can you get them out?
8. What effect is the fire having on the exposures?

Fire Command Instructor's Guide

Simulation Questions

Function #2
Situation Evaluation
IC Checklist

•Quickly identify and react to safety "red flags".

1. List any obvious red flags for the fire.
2. Do these red flags change your strategy and IAP?
3. List any potential red flags (unseen critical factors that have a high probability of existing).
4. Do these red flags change your strategy and IAP?
5. How do you find out about these unseen red flags?

■ Quickly identify and react to safety "red flags".

Red flags are the things that when you notice them you say to yourself, "Oh no." Red flags are information items that must be addressed because they can end up hurting/killing someone. Many of the major red flags are the regular pieces of the standard information inventory, listed above. These are information items that we routinely go after (checking the attic for fire extension, making sure the fire hasn't dropped power lines, etc.).

The IC must always use a pessimistic approach when sizing-up, assuming worse until proven otherwise. Many of the assignments the IC makes will be to verify areas that could be a problem. A red flag will not necessarily change the overall incident strategy or incident action plan but it must be identified and dealt with. This is a big way that the IC makes sure that everyone goes home when the event is over.

Fire Command Instructor's Guide

Situation Evaluation
Bullet point #2-9

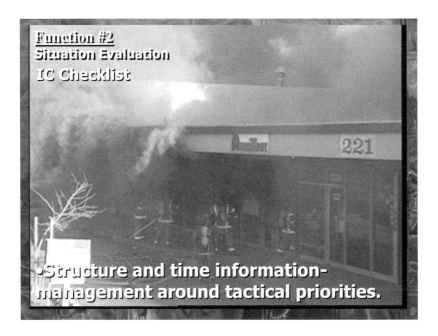

Function #2
Situation Evaluation
IC Checklist

•Structure and time information-management around tactical priorities.

■ Structure and time information-management around tactical priorities.

The tactical priorities are our job list for incident operations--they are the reason our customers call us. Since they represent the reason we are there, it only makes sense that we use these standard priorities as the basis for service delivery. It also provides a simple and understandable language for all the incident players. Assigning Engine 1 to "search the exposure to the west" is very definitive. Checking back with them after a proper length of time and asking if "they have and all clear in the west exposure" should lead to a simple and understandable "yes or no."

For the most part, effective incident communications are centered around completing the tactical priorities, within the parameters of the critical incident factors and firefighter safety.

Simulation Questions

1. What do the sector reports (progress reports) sound like for all the assigned sector positions at the incident scene?
2. Based on these reports, does the IC know when the tactical priorities have been completed in each sector?

Situation Evaluation
Bullet point #2-10

Simulation Questions

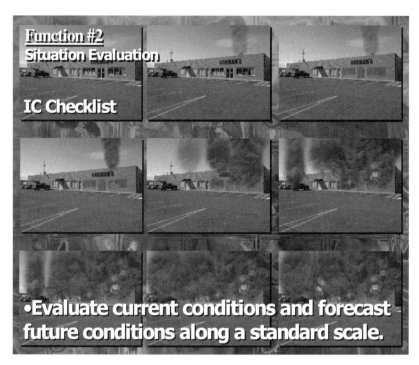

1. Where in the standard scale is our simulated fire?
2. Is the scale one of the main factors in determining:
   a. The strategy
   b. The IAP
   c. If we are operating within the risk management plan
   d. Required resources
   e. Forecasted conditions

■ Evaluate current conditions and forecast future conditions along a standard scale.

This standard scale shows the progression of the fire from "nothing showing" to "burned to the ground" (page 119 of the text book). Identifying the ten phases that separate the extreme ends of the scale (one and ten) provides a time line/fire progression that helps makes sense of how we plan to deal with the incident problem (the fire). When we look at the scale (in the book) the first three to four phases can be controlled with a well placed interior, offensive attack. As the scale moves to five and six, the fire is starting to involve major pieces of the structure, generating more of the nasty products of combustion. These middle stages make offensive operations a more difficult and dangerous operation. When we plug the risk management plan into the scale, phase number six makes offensive operations questionable and anything beyond it is a large offensive risk for no gain (the fire area is unsurvivable for any unprotected occupants and the structure is beyond saving).

When we apply the scale to fires in larger structures, the IC has got to factor in not only the time it takes to get the required number of attack lines to the seat of the fire (deeper inside a

Fire Command Instructor's Guide

Situation Evaluation
Bullet point #2-10

Simulation Questions

larger structure) but also how much time it takes to provide back-up to reinforce those positions, the length of time it takes to get resources to key safety positions (RIC Teams, Safety Sectors, etc.) and how long it is going to take to deliver the extra support required for larger structures (ventilation, forcible entry, hardening the exits/entrances, etc.). A key safety factor is how long it will take to move crews and personnel out of interior positions (particularly in larger structures) if conditions deteriorate and a change to defensive strategy is called for.

Effective operations in larger commercial buildings requires a much larger commitment of resources in the beginning of the event. The IC (along with the rest of the organization) cannot use the same approach to manage a fire in a 20,000 square-foot building, as s/he does in a 2000 square-foot house.

Simulation Questions

1. Does the IC know where resources are operating (assigned to our simulated fire)?
2. How does the IC stay connected to resources?
3. How does the IC keep from being unpleasantly surprised by changing conditions?

Function #2
Situation Evaluation
IC Checklist

•Continually reconsider conditions – stay current and stay connected to resources.

■ Continually reconsider conditions--stay current and stay connected to resources.

The IC's strategy and IAP are based on the initial size-up of incident critical factors. These critical factors are very dynamic--they are either getting better or they are getting worse (they never stay the same). This is one of the joys of our job--instant feedback. If the conditions are getting worse the IC must quickly figure out why and react accordingly. This capability is part art, science and experience. The problem is if the IC sticks with a deteriorating situation long enough the problem can end up doing great harm to the troops. There are no time outs and it is impossible to uncollapse a building.

The incident conditions are the things that drive the strategy, the IAP and what we apply to our risk management plan. We make a huge investment in training, using and refining the systems and SOPs so when we get to the scene and actually deliver service, we can quickly adjust what we are doing based on the incident conditions.

Fire Command Instructor's Guide

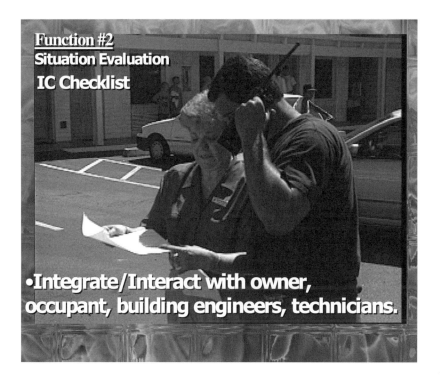

Function #2
Situation Evaluation
IC Checklist

•Integrate/Interact with owner, occupant, building engineers, technicians.

Simulation Questions

1. How many owners/ occupants will be effected by this fire?
2. Do you think that any of these "responsible parties" will have information that you can use tactically?
3. Is it possible that this information could have an effect on firefighter safety?
4. Do you assign someone to interact with these folks?

■ Integrate/interact with owner, occupant, building engineers, technicians.

These folks often times have the most knowledge about the incident problem. They know what the fuel load is comprised of, the layout of the building, and the different built in buildings systems that can help, or hinder, our efforts towards controlling the incident problem. Many times they also know what caused the incident problem and may have the most knowledge about the easiest and safest way of controlling the problem.

The owner/occupant are our customer. They should be sought out and consulted/briefed as soon as it is feasible to do so. In many cases there may be something that is of critical importance to them. If we can save or salvage these items it may be the difference between our customer staying in or going out of business.

Fire Command Instructor's Guide

Simulation Questions

Function #2
Situation Evaluation
IC Checklist

•Maintain a realistic awareness of elapsed incident time.

1. Does the IC factor in our pre-arrival time (the length of time the fire has been burning prior to our arrival)?
2. If you chose the offensive strategy, and committed crews to the interior of the fire occupancy, how much time will you give them to operate in this strategy?
3. How do you track incident time?
4. How does the IC track the length of time that each unit/ member has been on air (breathing out of their SCBA)?

■ Maintain a realistic awareness of elapsed incident time.

Time can do funny things at the incident scene. Sometimes hours seem like minutes. Other times minutes seem like hours. One constant for structure fires is that the building will hold up a very short period of time when it is exposed to fire. Another unforgiving time constraint is the length of time an SCBA's air supply will provide it's wearer.

In many systems the Communications Center provides the IC with elapsed time reminders. These reminders serve as triggers for the IC to re-evaluate conditions, the strategy and the length of time firefighters have been operating in the hazard zone as an example of this time management process, a 20-minute elapsed time notification should cause something to happen, if the IC has crews deployed to the interior of burning building where the fire is still out of control.

Another piece of elapsed incident time that we don't talk about much is the burn time from ignition to our arrival. We will rarely know the exact length of time the fire has been burning prior to our arrival but we can come pretty close to estimating with a good size-up of the conditions on our arrival. The 20-minute rule (if the fire has been burning uncontrolled for 20 minutes abandon the structure) that the fire service has used for decades must include the burn time prior to our arrival. As a note of interest (and survival) for many "modern" (i.e., light weight) structures, the 20-minute rule is quite generous.

Fire Command Instructor's Guide

Situation Evaluation
Bullet point #2-14

Function #2
Situation Evaluation
IC Checklist

•Consider fixed factors and manage variable factors.

Simulation Questions

1. What are the fixed factors for our fire?
2. What are the variable factors?
3. Can the IC cause the variable factors to go away?

■ Consider fixed factors and manage variable factors.

Fixed factors pertain to things that we can't change. Things like the way a building sits on a piece of property (although if the building burns to the ground we assisted with the change). If a building is bordered on two sides by railroad tracks the IC has very little chance in magically making those sides accessible for engine companies. If high tension electrical lines border four sides of a structure, ladder companies will have a much more difficult time accessing the roof or flowing elevated streams. If crews advancing hand lines into a burning structure have to go around ten corners to get to the seat of the fire it will be slow going and dangerous. The realities that these fixed factors present must be plugged into the IC's IAP. If something that normally takes three minutes to do suddenly is going to take twenty minutes because of a fixed factor, the IC must react, plan, and manage accordingly. Fixed factors exist before we get to the scene (most of them exist when the building was designed and built). Some fixed factors are so overwhelming that it creates too high a hazard to conduct interior operations (out-of-control fire in the fireworks factory). Other fixed factors, like well designed and working sprinkler systems, make doing our job safer.

Fire Command Instructor's Guide

Situation Evaluation
Bullet point #2-14

Simulation Questions

Variable factors are the things the IC has the ability to change--if a building is full of smoke, the IC can order ventilation. If the building is heavily secured, a ladder company can be assigned to open it up (force entry). Engine crews manage the nastiest variable factor, the fire, by applying a sufficient amount of water on it, overcoming the heat it is generating and putting it out. When we don't (or can't) control the variable factors, we want to be in safe locations (out and away) when the fire consumes enough of the building, causing it to fall down.

Fire Command Instructor's Guide

Function #3
Communications

Major Goal

To initiate, maintain, and control effective incident communications.

IC Checklist

•Use communications SOPs.

1. Describe the communications SOPs and plan that you will use for this incident.
2. Does the plan make the IC the central communications player?

■ Use communications SOPs.

Communications SOPs set the stage for how we are going to communicate with one another before the event occurs. They provide how we exchange information and describe how we use our hardware (radios, computers, mouths, ears, etc.) to get the job done. These SOPs are the ahead-of-the-event communications agreement and become the template we use to train, operate at the incident scene, evaluate, and revise the way we communicate.

Communications SOPs must be designed to prevent us from going into communications "overload." We need to develop systems and methods that allow the timely sharing of information, particularly when the incident has expanding and large resource requirements. At these types of incidents airtime on the tactical radio frequencies becomes more and more scarce. As the incident expands, more incident players are assigned and working and they all have radios (in many systems). If we don't develop a plan ahead of the event, the IC can easily get knocked of the air.

Fire Command Instructor's Guide

Simulation Questions

Function #3
Communications
IC Checklist

•Start/control communications upon arrival with initial radio report that describes conditions and actions

1. What is your initial onscene report for this fire?
2. Does it identify the incident problem, the chosen strategy, the initial IAP, and identify that an IC has been established for the incident?
3. Does the initial radio report put the IC in a position where s/he can control incident operations?
4. What SOPs automatically go into effect after the initial radio report has been given?

■ Start/control communications upon arrival with initial radio report that describes conditions and actions.

The initial radio report should set off a chain reaction of standard organizational elements. First and foremost, it lets the entire response (and anyone else listening to the radio) know that someone has arrived to the scene. The incident problem gets identifies and the strategy and incident action plan are declared. It also puts an IC in command of the event (i.e., assumption/confirmation). Command and communications SOPs are designed with the IC as the central communications player. If the IC can't control incident communications, s/he cannot control/manage the incident players.

A good initial radio report puts everyone who is involved in the incident on the same page. The first-due ladder may be a mile from the scene, but based on the initial report, chances are they have a pretty good idea of what action they will be taking when they arrive to the scene. Later arriving engine companies know to assume a staged position. Standard staging procedures let the IC know that the company has arrived and are in uncommitted positions, waiting for an assignment. This facilitates the entire communications (and management) process and allows the IC to communicate assignments (orders) based on his/her plan.

Fire Command Instructor's Guide

Function #3
Communications
IC Checklist

•Use effective communications
techniques to keep everyone connected.

Simulation Questions

1. Describe how the IC uses communications to keep the command staff (i.e., command team and sections) connected.
2. Describe how the IC keeps everyone else connected.
3. Describe how sector officers keep all of their assigned resources connected.
4. How do sector officers connect with other sector officers?
5. How do companies stay connected to their assigned sector officer
6. How do companies stay together (how do the four members of E-1 stay connected and together)?

■ Use effective communications techniques to keep everyone connected.

The IC uses the radio to 'link' all the incident players together. The end result should be the coordinated action of all the workers on all three of the organizational levels (strategic, tactical and task). The IC uses the tactical radio channel to assign units according to the IAP. This is why it is so important to have a plan (brain engaged) before you start talking (mouth engaged).

Not everyone can talk at once. The IC has to keep control of the airwaves. Lonely engineers, chiefs responding from twenty miles out, or anyone else that fills their void by chatting over the tactical channel can completely destroy the IC's ability to manage the incident. The IC must not let these "Chatty Kathy's" knock him off the air.

The IC connects everyone to the strategy when he/she announces the chosen strategy and then assigns and manages resources within an IAP that matches the strategy. This eliminates strategic confusion. As the IC assigns sector officer responsibilities the plan for that sector is communicated. This connects the strategic and tactical levels. Companies are assigned to do tasks within the sectors based on this plan, connecting all three levels.

Fire Command Instructor's Guide

Communications
Bullet point #3-4

Simulation Questions

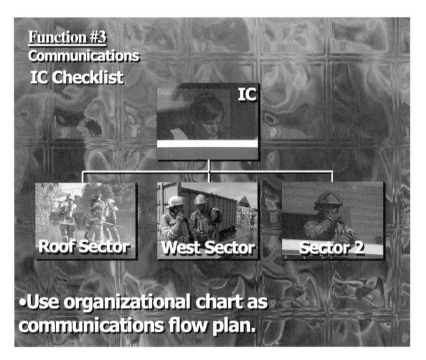

Function #3
Communications
IC Checklist

•Use organizational chart as communications flow plan.

1. Describe how the IC uses organization to streamline the radio communications process.
2. Describe the communications process/forms that sector officers use.
3. Describe effective communications within a sector--how do companies communicate with one another when they are operating in the hazard zone.

■ Use organization chart as communications flow plan.

Sectorizing the incident scene has a positive and profound effect on the communications process. When the IC assigns Sector Officer responsibilities to the officers initially assigned to the different key tactical positions around the incident site, it enhances the entire communications process. Units and personnel assigned to the sector usually communicate face to face with the sector officer (depending where everyone is physically located). The sector officer communicates with the IC over the tactical channel. In cases where the IC hasn't implemented/assigned sector officer assignments, s/he communicates with each individual unit assigned to the incident scene. As an example, if the IC has six different units assigned to the incident s/he communicates with each of those units over the tactical channel. If the IC can assign those six companies to three different sectors s/he cuts their communications partners in half. This frees up time and communications space for the IC. Free time that can be used to do the other seven functions of command.

Sector officers will communicate with one another over the radio or face to face depending on their proximity to one another. As the organization expands branch officers and section chiefs use the same communications model and additional radio frequencies (in some cases) to stay connected.

Fire Command Instructor's Guide

Function #3
Communications
IC Checklist

•Use companies, sectors, and alarm as communications partners.

Simulation Questions

1. Describe how the IC uses communications from companies and sectors to keep the strategy current and drive the IAP.
2. Describe how the IC uses these communications partners to provide for firefighter safety.

■ Use companies, sectors, and alarm as communications partners.

As the IC begins incident operations, s/he will base actions on the known (usually visual) conditions. This is what drives the strategy and the IAP. Keeping the strategy and IAP current requires filling in all the critical information holes. The IC does this by assigning units to key tactical positions. The companies assigned to these places take the appropriate action (within the strategy and IAP and based the conditions in their area). They report back to the IC about the conditions, their actions and any needs. This is the best use of communications--keeping everyone connected in order to solve the incident problem, and keeping everyone safe by basing our actions on the conditions.

Simulation Questions

Function #3
Communications
IC Checklist

•Maintain a clear, controlled, well-timed radio voice.

1. Describe the organizational effect that a screaming and frantic IC (over the tactical radio channel) has on incident operations.
2. Describe the organizational effect that a calm and collected IC (over the tactical radio channel) has on incident operations.
3. Which is better and who do you want to be?

■ Maintain a clear, controlled, well-timed radio voice.

The biggest evidence that an IC is in place and operating is to listen to how incident communications sound over the radio. When the IC is calm, cool, and collected, it spills over to all the incident participants. The other options include to scream, cry, stutter, or babble into the radio. None of these options instill much confidence in the troops.

Progress reports, orders and all other communications are more clearly understood when they are delivered in a steady, controlled voice. This applies to anyone that uses the radio (or any other means) to communicate.

Fire Command Instructor's Guide

Communications
Bullet point #3-7

Function #3
Communications
IC Checklist

•Listen critically -- understand communications difficulties from tough operating positions.

Simulation Questions

1. What is the worst communications position at our fire?
2. How do we balance/ overcome the communications problems associated with companies that must operate in the hazard zone?
3. What is the safety effect of having the IC operate in the hazard zone in a fast-attack position?
4. What is the safety effect of having the IC operate in a strategic position (supported, inside a command post)?

■ Listen critically--understand communications difficulties from tough operating positions.

The reason we put an IC in a strategic command post (outside the hazard zone in a vehicle designed to be a command post) is to be in the most ideal position to communicate (send and receive). Companies that are operating in the hazard zone are in the worst positions on the planet to try to communicate. Bundled up in sixty pounds of protective gear, breathing through a SCBA, and inserted inside a burning building makes it pretty difficult to carry on an intelligent conversation (over a five-watt walkie-talkie radio).

There are a lot of hazard-zone distractions that can cause communications problems. The IC needs to understand this when communicating with operating companies. Companies also must understand that their only communications link to the outside world is over their portable radio. The command system is dependent on the ICs and the operating units always being able to talk to one another.

Fire Command Instructor's Guide

Simulation Questions

Function #3
Communications
IC Checklist

•Mix and match forms of communications (face to face - radio - computers - SOPs).

1. What forms of communication will be utilized in the hazard zone of our fire?
2. What forms of communications will be used in the command post?

■ Mix and match forms of communications (face to face/radio/ computers/SOPs).

Communications is about sharing information. The incidents we respond to tend to be of a short, intense duration. Our communications systems, and our brains, are designed to share a given amount of info in a given period of time. The most used communications avenue (between each of the organizational levels) is the tactical radio channel. This provides a link to everyone that has a radio (which in many departments is everyone). The problem is that not everybody can talk on the radio at the same time. This makes SOPs, computers, real-time video and face-to-face communications very important elements of our communications systems.

Operations within sectors should be done face to face as much as possible. SOPs are powerful because they define what a task, or series of tasks are between all the participants--as an example, the IC only needs to order a ladder company to the roof to ventilate and assume roof sector. The responsibilities of the roof sector, what they need to include in their progress reports, and how to ventilate the roof are things that are practiced, trained on, applied, critiqued, and kept in the form of SOPs.

Fire Command Instructor's Guide

Simulation Questions

Computers are used for a whole host of communications and information management items--preplans, accountability rosters, maps, etc. Some departments are using enhanced video equipment to take real-time video from news helicopters and display it in the command post. If the pilots of these "eyes in the sky" have radio communications with the command post, the IC can request specific views of the incident site.

The goal is to use and balance all the communications system elements, without overtaxing them (particularly the tactical radio channel), to effectively manage the incident.

Simulation Questions

1. What kind of progress report would you give concerning the interior of the fire occupancy (based from the picture)?
2. What kind of progress report would you give concerning the exposures on either side of the fire occupancy (based from the picture)?
3. What kind of progress report would you give concerning the roof of the fire occupancy (based from the picture)?
4. How often do you (IC) want progress reports?
5. What information do you want the progress report to contain?

Function #3
Communications
IC Checklist

•Coordinate timely progress reports.

■ Coordinate timely progress reports.

The IC uses progress reports to keep the strategy and IAP current and to monitor the progress of work (completion of the tactical priorities). Progress reports should be structured around the status of the completion of the tactical priorities. If all the sectors have reported an "all clear," (completion of the primary search), that tactical priority can be checked off of the IC's tactical worksheet and efforts and resources move on to the next priority (fire control).

Any urgent information that pertains to safety or the completion of the tactical priorities should be immediately shared with the IC and anyone else who will be effected by the information. Example--"roof sector to Command, be advised there is heavy fire in the attic. The roof is unsafe, were coming down. Get everyone out from under the roof."

Fire Command Instructor's Guide

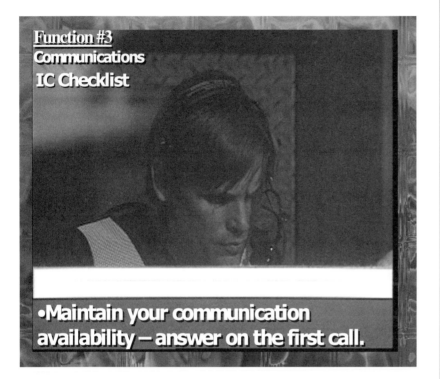

Function #3
Communications
IC Checklist

•Maintain your communication availability – answer on the first call.

Simulation Questions

■ Maintain your communications availability--answer on the first call.

A couple bullets back we talked about the IC understanding that operating companies are in tough communications positions. The IC shouldn't be (when they are operating in the command mode). The IC's main focus must be placed squarely on the units that are operating in the hazard zone. This is how we manage strategic level safety. The IC controls the strategy, manages the IAP and serves as the link to the outside world (and help) to all the workers that are operating in the hazard zone. The tool the IC uses to pull all this off is the radio. The IC must always operate the system (build, expand, reinforce...) in such a way that allows him/her to stay connected to the folks who are operating in hazardous positions.

The IC does this by always being available and able to communicate, particularly with the folks operating in the hazard zone.

1. List the consequences when the IC doesn't (for whatever reason) answer the radio:
   a. Creates unsafe situation--no one managing strategic level safety.
   b. Crews may end up freelancing--in absence of an IC crews will take independent action
   c. The strategy and IAP may get out of whack.

Simulation Questions

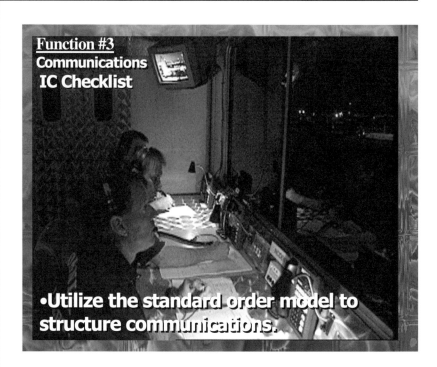

•Utilize the standard order model to structure communications.

Communications
Bullet point #3-11

1. Describe the order model you use to make sure that the receiver understands the sender.

■ Utilize the standard order model to structure communications.

The order model is used to keep communications simple and to verify that orders were understood. Example--the IC has two engine companies conducting and interior attack inside a medium-sized store. Both companies entered from the north and are pushing fire out of the windows on the south side of the store (just like it's supposed to happen). The IC orders the third engine company to lay a supply line to the south and take an attack line into the west exposure for search and rescue and to verify that the fire hasn't extended. The officer of the third engine comes back with "Engine 3 copy, laying a line and attacking the fire from the south." Because Engine 3 followed the order model (parroting back the order), the IC knows that Engine 3 will seriously disrupt the fire attack if they do what they said they were going to do (and what they really want to do). The IC comes back and says "NO, I want you to check the exposure to the west..." This goes on until Engine 3 gets it right. "10-4" doesn't cut it.

Communications
Bullet point #3-12

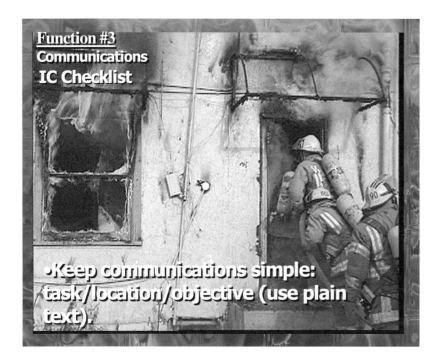

Function #3
Communications
IC Checklist

•Keep communications simple: task/location/objective (use plain text).

■ Keep communications simple: task/location/objective (use plain text).

Incident operations are conducted around the completion of the tactical priorities. Incident communications should mirror this simple concept. When the IC assigns companies based on a well thought out IAP, everything seems to naturally fall into place and companies will base their progress reports on the original orders the IC gave them. Life is wonderful.

When the IC assigns companies to tactical positions they should have a clear and definitive assignment that fits into the IAP. Ordering Engine 1 to "come on in and give us a hand" conveys the message, "come into the scene and do what ever you want because I really don't know what I want you to do."

1. List the orders that the IC will give to the first two engine companies and the first ladder company.
2. Are these orders a natural outgrowth of the strategy, IAP and incident conditions?

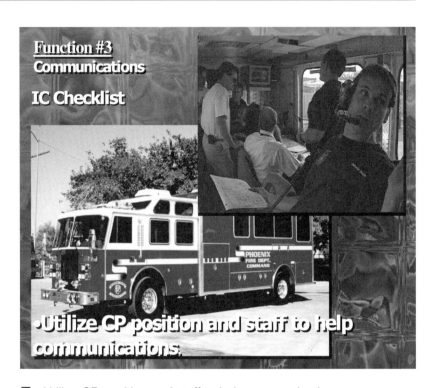

Function #3
Communications

IC Checklist

•Utilize CP position and staff to help communications.

1. Describe how a command post improves the IC's communications capability.
2. Describe how a chief's aide improves the IC's communications capability.

■ Utilize CP position and staff to help communications.

The IC must be able to control the pace of incident operations. When an incident continues to escalate and more companies are deployed into the hazard zone the IC must be able to keep up. S/he does this by expanding the command organization. In many cases larger incidents are managed over multiple channels. It is a serious distraction to have to listen to more than one radio channel at a time.

This escalation of command must be a normal occurring progression that keeps pace with the incident. When the IC has a support officer (the beginning of the command team) the SO can fill out and keep the tactical worksheet current. This frees the IC to focus his/her attention and energy on sizing up conditions and communicating with hazard zone resources. Putting staged (level two) and responding resources on another radio channel frees up more airtime on the tactical channel and eliminates scores of uncommitted resources that the IC would otherwise have to communicate with and around. Implementing this channel requires another person in the incident management food chain to communicate with and manage those resources. These positions (staging or logistics in this case) are worth their weight in gold because they allow the IC to manage the hazard zone.

Fire Command Instructor's Guide

Simulation Questions

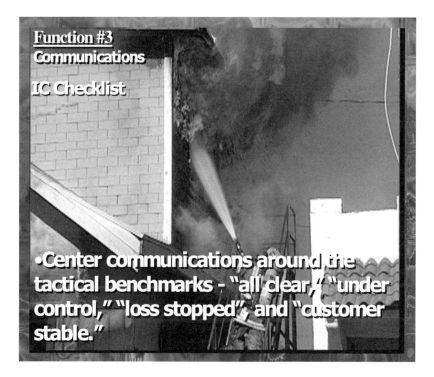

Function #3
Communications

IC Checklist

•Center communications around the
tactical benchmarks - "all clear," "under
control," "loss stopped", and "customer
stable."

■ Center communications around the tactical benchmarks--"all
clear", "under control", "loss stopped" and "customer stable."

Communications should center around the completion of the
tactical priorities and firefighter safety. Sector officers and
companies should base their communications on conditions
in their assigned area, action taken and the effect it is having,
and any needs they have. This will help keep communications short, to the point and effective. It also maximizes the
available free air time, making it available for important
messages.

Incident operations are conducted around the completion of
the tactical priorities. Incident communications should mirror
this simple concept. When the IC assigns companies based
on a well thought out IAP, everything seems to naturally fall
into place and companies will base their progress reports on
the original orders the IC gave them. This keeps the operation focused on what we showed up to do--make sure
everyone is okay, elimination of the incident problem, reduce
the harm to the customers stuff.

1. What are some of
   the information
   items that the IC
   will base his/her
   communications
   on?

Communications
Bullet point #3-15

Function #3
Communications

IC Checklist

•Project a good radio image

1. List the things that assist the IC to project a good radio image.

■ Project a good radio image.

People have confidence in a competent and calm IC. One of the main ingredients that go with being an effective IC is how well you use the radio. The IC has the powerful ability to pace (slow down) the companies' activity meters just by the way s/he talks over the tactical channel. Things run smoother and more effectively when the event is not an emergency to us. Compare this to the frantic IC that screams and blabbers over the radio. This tends to be very contagious--companies will start mirroring the behavior of the boss and before you know it, the incident looks (and sournds) like the Keystone Cops. This is bad. Stay in control and you remain in charge. I don't know who said that, so I am going to take credit for it.

Fire Command Instructor's Guide

Deployment

Bullet point #4-1

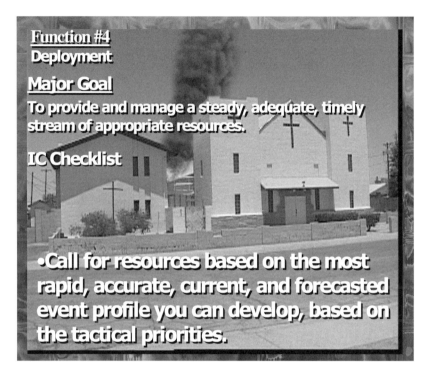

Function #4
Deployment

Major Goal

To provide and manage a steady, adequate, timely stream of appropriate resources.

IC Checklist

•**Call for resources based on the most rapid, accurate, current, and forecasted event profile you can develop, based on the tactical priorities.**

Simulation Questions

1. How many resources will be required for our fire?
2. How long will it take to get them there?
3. Does the IC base initial and ongoing resource requests on completing all four of the tactical priorities:
   a. Operational units-- engines, ladders, squads, etc.
   b. Command officers
   c. Support units-- utility trucks, rehab units, etc.
   d. Functional units-- fire investigators, occupant/cus- tomer support advocates--Red Cross, etc.

■ Call for resources based on the most rapid, accurate, current, and forecasted event profile you can develop, based on the tactical priorities.

The IC manages the tactical and task level of the organization in order to achieve the tactical priorities. The last thing the IC (or anyone else) wants or needs is to run out of resources before incident operations have been safely concluded. It causes a very sick and lonely feeling when the IC discovers important tactical work must be done but there are no resources (available and on the scene) left to do it.

Part of the IC's initial size-up process involves figuring out how many companies/units/equipment/personnel/specialized help it will take to do (and finish) the job and make sure that help is requested and on the way. The experienced IC uses a combination of past experience tempered with a pessimistic approach (it takes more, not less) when calculating the resource level required to conduct incident operations.

Fire Command Instructor's Guide

Simulation Questions

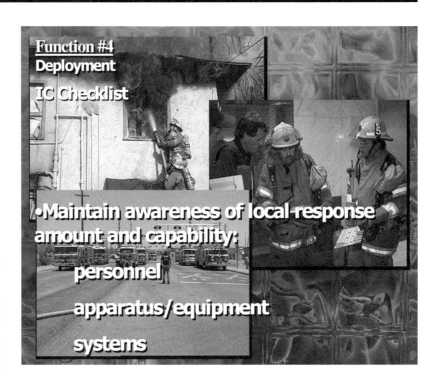

Function #4
Deployment
IC Checklist

•Maintain awareness of local response amount and capability:

personnel

apparatus/equipment

systems

1. List all the different response agencies that will be required for our fire.

2. What systems does the IC use to get all of these different agencies to the incident scene and then to get them all to work together towards the same common goals?

■ Maintain an awareness of local-resource response, amount, and capability:
   • personnel
   • apparatus/equipment
   • systems

The IC is the person that has to match (and manage) the work that must take place at the incident scene to the people and equipment that will be doing the work. Matching these two constants (task and workers) requires that the IC have a good grasp of the available area personnel, equipment, apparatus and the systems used to activate and manage those resources.

We are not the lone emergency service providers. It is not uncommon for five or more different agencies to respond to Mrs. Smith's house fire (local fire department, law enforcement, power/utility company, Red Cross, etc.). The IC must include the participation of other incident players and agencies and incorporate their involvement as a regular part of everyday operations.

Deployment
Bullet point #4-3

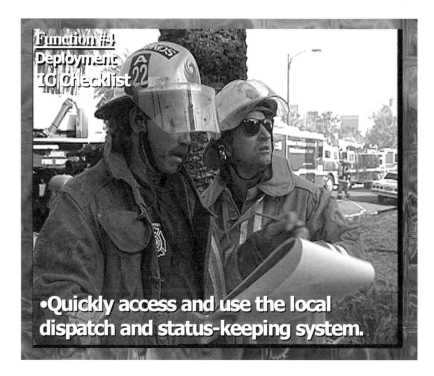

Function #4
Deployment
IC Checklist

•Quickly access and use the local dispatch and status-keeping system.

1. How many re-sources (alarms) will you use for this fire?
2. How does the dispatch center fill this request-- mutual/automatic aid, standard alarms/assign-ments?
3. How does the IC, sector officer, company officer track assigned resources?

■ Quickly assess and use the local dispatch and status-keeping system.

The IC requests resources from the dispatch center. As re-sources request are made, the dispatch center figures out what resources are available and fills the request by sending the closest, most appropriate resources. Making resource allocation a function of the communications center streamlines getting the right type and amount of resources to the scene. If the IC needs eight engine companies, the power company, local law enforce-ment, and the Little Sisters of the Poor, s/he doesn't really care where they come from, so long as they end up at the incident scene in a timely fashion.

Mutual and automatic aid from neighboring departments is effective and has a positive impact on service delivery when it happens seamlessly. If the communications center has to call the mayor from their city before they call the mayor from the mutual aid city, the closest resources should show up sometime next month, after several rounds of meeting at each mayor's office.

Fire Command Instructor's Guide

Deployment
Bullet point #4-3

Simulation Questions

The communications center sends the IC the resources s/he requested and they maintain service delivery for the rest of the community. If the IC is working a "big deal" event that involves greater alarms, the communications center is busy moving available resources to fill any response gaps created by the big event.

The IC is responsible for keeping track of all of his/her assigned resources. S/he uses a tactical worksheet to do this. The dispatch center is responsible for keeping track of all of the resource they dispatch. They use everything from 3" X 5" cards to satellites to accomplish this mission.

**Deployment
Bullet point #4-4**

Function #4
Deployment

IC Checklist

•Monitor and manage within on-line response times.

1. How long will it take to get the required resources to the scene--use a location in your own jurisdiction that has similar structures?
2. Can these times end up effecting the IC's strategy and IAP? How/why?
3. Does the IC also factor in the length of time it will take to get resources in place and operating after they have been given orders/assignment by the IC?

■ Monitor and manage within on-line response times.

This includes matching the time it takes to get the right amount of resources (people and stuff) into the area where the work must be done (the incident site).

This is one of the constants that falls under deployment--when the IC sizes up a structure fire (as an example) and determines that it is going to require three engine companies and two ladder companies to conduct a successful offensive firefight. Next the IC figures in the constraints of time and estimates that those resources must all be in place (inside and operating) within the next three minutes. The IC must match tactical/task action on the size up of the incident problem at that particular point in time.

Fire Command Instructor's Guide

Simulation Questions

1. Why are staging absolutely critical for the IC's capability to effectively manage incident scene operations?

2. Describe the standard work cycle for the first two engine companies and the first ladder company assigned to this incident.

3. Explain the effect that rehab sector has on the standard fire fighting work cycle.

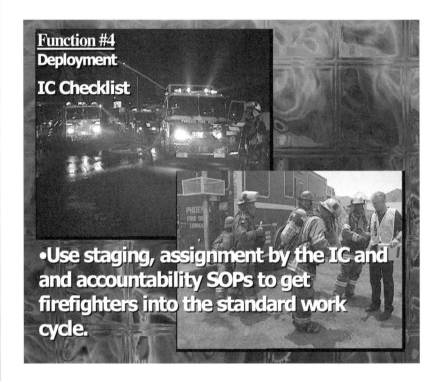

Function #4
Deployment
IC Checklist

•Use staging, assignment by the IC and and accountability SOPs to get firefighters into the standard work cycle.

■ Use staging, assignment by the IC and accountability SOPs to get firefighters into the standard work cycle.

The IC is the resource allocator for the incident, and is responsible for managing all assigned resources work cycles on the strategic level. The standard work cycle looks like this:
- dispatched
- responding
- staged
- assigned to incident by the IC
- working
- rehab
- ready for reassignment
- available

This creates a resource delivery system that allows the IC to deploy resources according to his/her IAP and provides a system

Deployment
Bullet point #4-5

Simulation Questions

that allows the IC to manage accountability on the strategic level. If companies do not stage when they get to the scene they will not be assigned according to the IC's plan. They will auto-assign themselves according to their own freelancing-based reactions. Freelancing will cause any incident management system to fail.

Once assigned to operational positions, units operate within the accountability SOPs. The foundation of accountability SOPs is the constant that crews stay together and always maintain the ability to exit the hazard zone.

Firefighting is very strenuous work. If the IC doesn't provide for the needed rehab, the fire incident can quickly turn into a medical incident--hauling over-worked, heat-stressed firefighters to the hospital. It isn't very nice (or very effective) when the IC does this, so rehab must be a regular, standard piece of incident operations. A proper rehab sector allows the IC to put assigned resources through the standard work cycle several times (assigned, working, rehabbing, assigned, working, rehabbing, etc.).

Simulation Questions

1. Fill out a tactical worksheet for this incident.
2. Based on the tactical worksheet, can the IC:
   a. track/control the position and function of all assigned resources
   b. manage the completion of the tactical priorities
   c. ensure that all critical areas/ functions have been addressed.

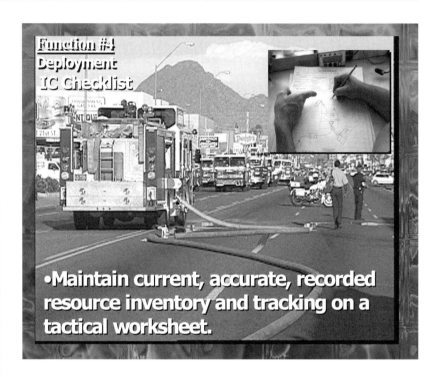

Function #4
Deployment
IC Checklist

•Maintain current, accurate, recorded resource inventory and tracking on a tactical worksheet.

■ Maintain current, accurate, recorded resource inventory and tracking on a tactical worksheet.

The best way to remember something is to write it down. This is particularly true when you're managing an ongoing event, as dynamic and dangerous as a fire. The tactical worksheet provides the IC with a mechanism to track where people are and what they are doing and who is their boss.

As the IC makes assignments, s/he should note them on the tactical worksheet. The IC/command team must manage the tactical worksheet in such a way that it allows them to know the general location and assignment of all assigned incident resources. Managing the tactical worksheet in this fashion is the practical basis of strategic-level accountability.

Fire Command Instructor's Guide

Function #4
Deployment
IC Checklist

•Balance resources with task (don't overmatch).

Simulation Questions

1. How many units will be required to complete the tactical priorities in the fire occupancy?
2. How much time will it take?
3. How many units will be required to complete the tactical priorities in the exposures?
4. How much time will it take?

■ Balance resources with task (don't overmatch).

It doesn't make any sense to order a company to do more work than is realistic. In fact it can be very dangerous. Firefighters are very good at following orders and will attempt to carry out any fire fighting task the IC assigns them. When the IC orders an ladder company to, "Vent the roof, secure the utilities, provide lighting on the interior and get me a salvage size up," s/he has unrealistic expectations and needs to spend a couple of months working on a ladder company.

The IC must base assignments on the capabilities of the company/unit receiving the order. One of the constants, for folks that operate in the hazard zone, is the amount of time (and air) that an SCBA will provide. The IC cannot assign a thirty-minute task (that takes place in/around the products of combustion) to a group of workers that have a twenty-minute air supply. This is a very reckless approach to conducting incident operations and, when done over a period of time, will produce a number of injuries and fatalities that should be expected.

Simulation Questions

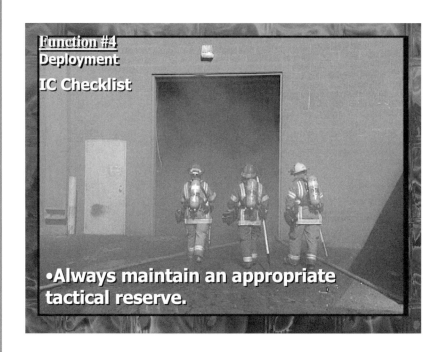

1. What kind of tactical reserve do you maintain for this fire?
2. Where will you position them?
3. How long does it take to deploy staged units to operational positions (how much time does it take for them to go from staging to where they are actually needed)?

■ Always maintain an appropriate tactical reserve.

Tactical reserves are a beautiful thing. The IC needs to resist the temptation to assign resources as soon as they report that they're staged. The wise IC will always hold a little back. These uncommitted resources can be used to fill any unexpected holes in the IAP or they can be utilized to assume RIC/RIT positions.

These uncommitted resources become the foundation for "plan B." They are also prevention for the queasy feeling the IC gets in the pit of his/her stomach when some urgent need suddenly presents itself and everyone else is already committed to incident operations.

Appropriate means not holding an alarm and a half in reserve for a simple room and contents fire in a ordinary, 1600 square foot, single family residence (house). It also means holding back a lot more than a single engine company for offensive fire fight in a large commercial structure.

Fire Command Instructor's Guide

Function #4
Deployment
IC Checklist

TACTICAL WORKSHEET

•Use command SOPs to manage and escalate operations.

Command

Simulation Questions

■ Use command SOPs to manage and escalate operations.

The IC controls the deployment of assigned resources by decentralizing the incident scene and making sector (division/group) assignments to where the work is actually taking place. Sector officers manage the deployment (i.e., the position and function) of their assigned resources. This provides a higher level of supervision, management, and all around safety for all the incident players, particularly the ones who are operating in the hazard zone. It also allows the IC to manage a rapidly escalating event and stay focused on the strategic level.

In many systems, when the IC request a greater alarm, level-two staging procedures automatically go into effect. This puts resource close to the scene, in a centralized location. Escalating events will also require more organizational support. These command/organizational resources will be needed to manage functional sectors (rehab, staging, etc.).

As the resource requirements required to bring the incident under control escalate, so must the IC's amount of IC support. This is where we implement command teams and (if the incident continues to expand) begin to fill out and assign section responsibilities. Supporting the IC in this fashion strengthens "Command" and keeps the IC ahead of the power curve.

1. How does the IC manage escalating operations?
   a. staging procedures
   b. rehab
   c. sector responsibility in key tactical positions
   d. IC support

Fire Command Instructor's Guide

Simulation Questions

1. Describe the risk management plan your department uses.
2. Where in the risk management does our fire occupancy fall?
3. How about the exposures?
4. Does the IC use the risk management plan throughout the incident or just in the beginning, when s/he is doing the initial size up?

Function #5
Identify Strategy/Develop Incident Action Plan

Major Goal
To use a regular, systematic method to make strategy decisions and to develop and initiate an incident action plan.

IC Checklist

•**Apply the standard risk-management plan throughout the incident.**

■ Apply the standard risk-management plan throughout the incident.

The risk-management plan not only describes the level of risk we will expose ourselves to, but also the reason (or gain) why we take that risk. The risk-management plan described in the text is in direct harmony and alignment with the tactical priorities, in fact, they are so closely matched, it almost takes on the look of the chicken or the egg riddle (which came first, the chicken or the egg?).

We must always factor in the effect the products of combustion are having on unprotected occupants. Our protective gear offers us a level of protection that in an atmosphere that would kill an unprotected person in seconds. If the IC is uneasy about the positions his/her troops are operating in because of the inordinate risk they are exposed to, it is time to take a serious risk-management plan reality check. As a general rule of thumb, incident operations will normally begin when the conditions are at their worst. Incident conditions should improve (become safer) after the initial attack is put into place. The ongoing evaluation of

Fire Command Instructor's Guide

Simulation Questions

operational effectiveness is a very big deal that directly impacts both service delivery and firefighter safety. Chapter seven is devoted entirely to evaluation, review, and revision. This process is based on completing the tactical priorities through the application of the risk-management plan to the critical incident factors.

It is impossible to write something in a book that the IC can refer to for every fire that s/he commands. This is a decision that rests with the IC and the rest of the command organization. They must apply a sane and lucid, risk-management plan to the incident conditions.

Fire Command Instructor's Guide

Strategy/Incident Action Plan
Bullet point #5-2

Simulation Questions

1. List the critical factors for our fire (we did this in function #2).
2. Based on these critical factors, what strategy will we employ?

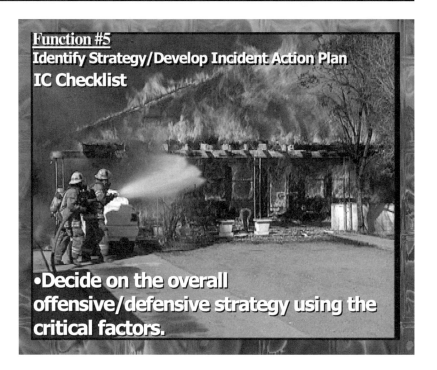

Function #5
Identify Strategy/Develop Incident Action Plan
IC Checklist

•Decide on the overall offensive/defensive strategy using the critical factors.

■ Decide on the overall offensive/defensive strategy using the critical factors.

The risk-management plan is the tool the IC uses to evaluate the standard action that needs to be applied to the critical factors and then to determine the proper strategy. If the critical factors add up to being able to conduct a primary search and to control the fire from interior positions, this leads us to an offensive operation. If the products of combustion have taken control of the structure, and none of the interior spaces are survivable, and all the property is essentially lost, the defensive strategy is the appropriate choice.

The IC must factor in the capabilities of the incident responders, particularly as they relate to the number of responders in relation to the size of the potential hazard, and the work that must be done to control that hazard, when determining the strategy. An offensive fire in a medium-sized single family residence is a

56

Fire Command Instructor's Guide

Simulation Questions

completely different animal than the same sized fire in a 30,000 square-foot commercial occupancy. Two or three engine companies and a single ladder company probably are enough resources to overpower the house fire. This same resource level may find themselves dangerously over-matched and in trouble within short order at the commercial fire. A key critical factor is the amount of fuel (building/contents) that isn't on fire. This relates back to travel distance/time required to reach the problem (with hose lines), how long it takes to get out of the hazard zone (once we get in), and the expanded number and size of concealed spaces where the fire can hide and suddenly rear it's ugly head.

Simulation Questions

1. Write out the initial radio report for this fire.
2. Describe the operational effect that declaring the strategy as part of the initial radio report has on incident operations.

Function #5
Identify Strategy/Develop Incident Action Plan
IC Checklist

•Declare the strategy as part of the initial radio report.

■ Declare the strategy as part of the initial radio report.

Declaring the strategy, up front, as part of the initial radio report, puts everyone on the same page. This eliminates any mystery of how we are going to operate at the scene, along with where we will do it (i.e., inside or outside the hazard zone). The strategy is what drives the tactics. If the IC declares the offensive strategy as part of the initial radio report, one would assume that exterior deck guns and other master stream devices are not going to be part of the tactical plan (not unless conditions change drastically). If the IC declares the defensive strategy, responding resources are thinking of big, outside water and cutoff points.

Including the strategy in the initial radio report also matches it to the description of the critical fireground (or incident) factors. Alarm bells should be going off throughout the system if the initial IC reports that they, "are on the scene of a large, abandoned warehouse, full involved," and finishes with "we are going in with an 1 ½" attack line for search, rescue, and fire control. We will be operating in the offensive strategy." Based on this initial report, the initial IC must be stopped. I have seen instances of this exact thing happening that were quickly remedied when the responding chief asked the initial IC (over the tactical radio channel) to re-evaluate the strategy. Having the IC declare the strategy should cause them to pause long enough to do an actual size up and then match their actions to those conditions.

Strategy/Incident Action Plan
Bullet point #5-4

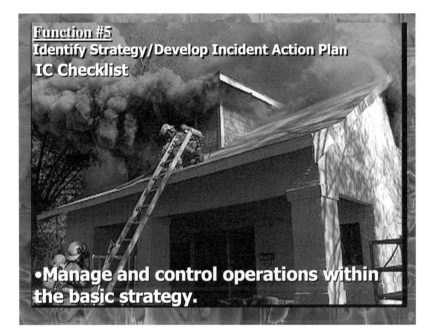

Function #5
Identify Strategy/Develop Incident Action Plan
IC Checklist

•Manage and control operations within the basic strategy.

Simulation Questions

■ Manage and control operations within the basic strategy.

The strategy defines the positions the troops will be operating from--offensive is inside; defensive is outside. This is one of the key ways the IC manages safety on the strategic level and completes the tactical priorities. The IC uses the strategy to get everyone moving in the right (i.e., safe, effective) direction and working together to solve the incident problem.

The IC also uses the strategy to shift the operation when the incident conditions change. Example: The IC arrived at the scene and sized up the fire as offensive. The IC formulated an IAP around the strategy and the critical incident factors. The fire has gotten progressively worse. The IC makes the decision that the plan (and strategy) that was developed five minutes ago is no longer effective. The IC declares the new strategy, with emergency traffic over the tactical channel, that s/he is changing the strategy to defensive. After operating companies report that they are safely out of the hazard zone and have a PAR a new plan of attack is formulated.

It is important to note that strategy comes first, then comes the IAP. In our example, the IC pulled everyone out of positions that were no longer safe to operate in. That was the most critical priority at the point in the operation. Only after everyone's safety is verified is the new IAP (the corresponding tactics) implemented.

1. Describe the relationship between the critical incident factors, the risk-management plan, the tactical priorities, the chosen strategy, and the development of the IAP.
2. Explain why the strategy drives the IAP and not the opposite.

Strategy/Incident Action Plan
Bullet point #5-5

Simulation Questions

Function #5
Identify Strategy/Develop Incident Action Plan
IC Checklist

•Implement an action plan to match the overall strategy.

1. What is your IAP for our fire?
2. Does it begin to address all the tactical priorities and provide for firefighter safety?
3. Doe the IAP change over the course of the incident?

■ Implement an incident action plan to match the overall strategy

The strategy drives the IAP. Most offensive fires are traditionally conducted around quick action that involves smaller diameter attack lines, a primary search and support work that addresses ventilation, forcible entry, utility control, providing access, etc. The strategy defines the goals we work toward (for offensive fires: all clear, under control, loss stopped, customer stabilized) along with where the workers will engage the incident problem (inside the hazard zone for offensive fires). The tactics describe the operational details of how we will achieve the tactical priorities. The tactics are translated in the form of orders. Example: "Command to Engine 1… I want you to lay a supply line to the north side and advance a handline to the interior. Keep the fire from extending to the north through the attic. Also do search and rescue in your immediate area. You will be North Sector".

The sum of these orders given to the various companies/sectors combine to form the IAP. After a while, the IAP for offensive fires starts to sound the same. This same thing also applies to IAPs for defensive fires. That is because the system we use to manage these events becomes more standard and routine the more we use it.

Strategy/Incident Action Plan
Bullet point #5-6

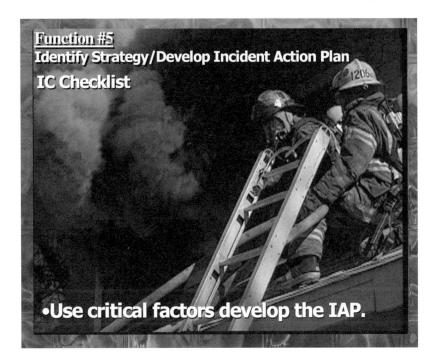

Function #5
Identify Strategy/Develop Incident Action Plan
IC Checklist

•Use critical factors develop the IAP.

■ Use critical factors to develop the IAP.

The critical factors are the set of incident conditions that causes one fire to be different from the next. We use a standard approach and incident management system to develop and conduct operations (our IAP) around the critical incident factors. Combining the critical factors with our risk-management plan is how we determine the proper strategy. The critical factors then become the things we attempt to change to solve the incident problem(s) (putting the fire out, controlling the products of combustion, along with the effect they are having on people and their things).

Critical factors are the conditions that effect our ability to achieve the tactical priorities. They are also the things that injure and kill us. The IC will begin operations knowing some of the critical factors. Other critical factors may be unknown. The IC develops the IAP based on the factors that must be addressed immediately along with the unknown factors that must be found out. One simple way to determine this is to ask the question, "Based on what is happening now, what is the worst thing that can happen next?" This combines both forecasting and pessimism (effective

Simulation Questions

1. Do all the known critical factors match up with your IAP?
2. Are there unknown critical factors that can change your strategy and IAP?
3. If there are unknown critical factors, how do you find them?
4. Is it possible to find out about these critical unknowns before they cause something bad to happen?

Simulation Questions

IC traits). Example: the IC sizes up the following critical factors-- light smoke from a commercial occupancy that is fifty-years old and has a large, heavy truss roof. The critical factors add up to the offensive strategy. The worst thing that can happen is that the fire is somewhere in the truss loft and if it burns up there for a sufficient amount of time the roof will collapse and kill everyone underneath it. The solution is to *quickly* determine/verify that the fire is not in the attic and has not effected the roof structure. If it is make the proper adjustment to the IAP.

Fire Command Instructor's Guide

Strategy/Incident Action Plan
Bullet point #5-7

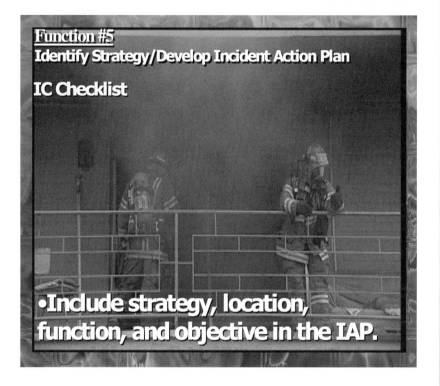

Function #5
Identify Strategy/Develop Incident Action Plan

IC Checklist

•Include strategy, location, function, and objective in the IAP.

Simulation Questions

1. What is the IAP for this incident?
2. List the IAPs for each sector.
3. Does the combination of all the sector IAPs = the IC's IAP?

■ Include strategy, location, function and objective in the IAP.

Incident action plans describe our operational plan for completing the tactical priorities. IAPs should be short and to the point. They also go out of date as conditions and priorities change. Starting out with the strategy defines the overall approach of the operation. The location identifies the areas and the corresponding key tactical positions where operations will be initiated. This process requires the IC to prioritizes the first, second, third, etc. areas that must be addressed. Operational functions define the tactics to be used in these locations. Functions include search and rescue, fire control, ventilation, roof operations, salvage, etc. Objectives are things that translate into the orders the IC uses to assign and get units into the attack plan--keep the fire from extending to the west exposure and get an all clear on the west exposure.

The IC should brief sector officers on the strategy and objectives that they are responsible for in their assigned area. The IC must have an "overall" IAP for the entire incident operation. Sector officer's IAPs should be much narrower in focus to describe the action in that sector and must fit into the overall incident operation.

Fire Command Instructor's Guide

Simulation Questions

1. What order will you complete the tactical priorities?
2. If you have chosen the offensive strategy in the fire occupancy will you flow any water prior to getting an "all clear?"
3. Describe the situations where we will do fire control activities in order to facilitate completing the primary search.
4. Describe how you will re-evaluate and adjust the strategy and IAP accordingly after completing the primary search and getting an "all clear."

Function #5
Identify Strategy/Develop Incident Action Plan

IC Checklist

•Use tactical priority benchmarks as action planning road map.

■ Use tactical priority benchmarks as the action-planning road map.

As we mentioned in the preceding bullet, IAPs go out of date as conditions and priorities change. When the IC gets completion reports from around the incident scene that one priority has been completed, it is time to move on to the next one. The IC keeps the plan current by continually matching action to conditions and focusing efforts towards the current tactical priority. This is the way that we manage something as dynamic as a structure fire (or any other emergency that we respond to).

The tactical priorities are the job list we use for incident operations. The tactical priorities are also the template for our risk management plan (or the other way around, depending on where you stand on that whole chicken or the egg thing). The IC may assign personnel to high-risk positions if search and rescue is a high priority (a decision the IC bases on the critical factors--a working fire in a single-family residence that has reports of people trapped, as an example).

64

Fire Command Instructor's Guide

Simulation Questions

Life safety is the only reason we should be taking a big risk. After life safety has been addressed the operation shifts to one where we are protecting property (by eliminating the incident problem). We will take a small and highly managed risk to achieve this tactical priority. Fire control and property conservation are closely tied together. A big reason we put the fire out is to make it stop damaging the property. A certain amount of damage will result from our fire control activities and are almost unavoidable. Most sane people do not want hose lines that discharge over one-hundred gallons of water per minute pulled into their homes or businesses. If those same home or businesses are on fire, it is a different matter. This does not give us carte blanche to do unnecessary damage under the guise of fire control. We must balance fire control efforts with the ultimate reason we are conducting them--to save the customer's property. This is in no way meant to imply that if we suspect that the fire is actively burning in some part of the structure that we wring our hands and hesitate opening up whatever part of the structure that we need to in order to access, find and kick the fire's ass. On the other hand, if we have the fire and products of combustion confined to the room of origin, we don't break every piece of glass out of the building.

Strategy/Incident Action Plan
Bullet point #5-9

Simulation Questions

Function #5
Identify Strategy/Develop Incident Action Plan
IC Checklist

•Re-announce ongoing strategy confirmation as part of elapsed time reports.

1. How often doe the IC re-announce the strategy?
2. How is the IC prompted/reminded to do this?

■ Re-announce ongoing strategy confirmation as part of elapsed time reports.

This should cause the IC to reevaluate if the strategy is still appropriate. Time is one of the basic critical factors. It is also one of the factors that we can't change. Elapsed time notifications are based on the time that the dispatch center received the initial call. It is rarely, if ever, based on the time that the fire actually started. Part of this initial size up should be a calculation of how long the fire was burning prior to our arrival. This is not always an easy thing to do, and it is often times a factor that the IC can find him/herself playing catch up with, which will eventually lead to nasty suprises. Depending on the conditions (where the fire is), the IC may have to get reports from multiple locations to get an accurate determination of burn time and fire extent/location.

This all takes place while the building is on fire. Buildings do not stand up very long when they host free-burning fires. Most buildings will not stay together for twenty minutes--our old traditional rule of thumb. I am sure you are as tired of reading it

Fire Command Instructor's Guide

Simulation Questions

as I am of writing it, but one more time, the IC's strategic decision is a compilation of the critical incident factors (which includes **total** burn time), the risk-management plan, and the tactical priorities. Save the customer, put the fire out, and protect the customer's stuff, but don't kill the troops in the process. If the fire has a big enough head start, there is no one to save and the property is lost.

Re-announcing the strategy also reminds all the incident participants of what strategy is being used to control the incident problem. These periodic announcements provide a very positive safety effect.

Fire Command Instructor's Guide

Simulation Questions

1. If you chose the defensive strategy for the fire occupancy, is it possible to assign units to offensive positions in the adjoining exposures?

2. If you choose to employ the strategy/tactics in question #1, how do you make everyone aware of the operation?

3. Describe why it is not a good practice to send attack crews into the rear of the fire occupancy and then put a master stream through the front windows.

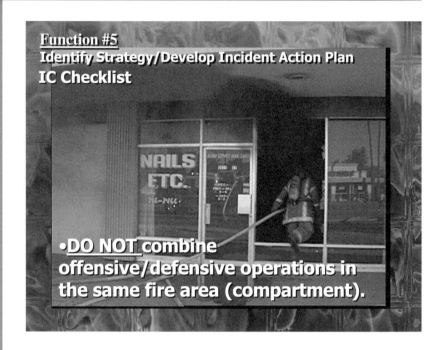

Function #5
Identify Strategy/Develop Incident Action Plan
IC Checklist

•DO NOT combine offensive/defensive operations in the same fire area (compartment).

■ Do not combine offensive/defensive operations in the same fire area (compartment).

Interior crews do not like it when come idiot introduces a master stream into the same area they are operating in. The IC is responsible for making sure this doesn't happen. When everyone operates within the system, these types of stupid mistakes are eliminated.

It is possible, and sometimes desirable and appropriate, to mix strategies on the fireground, so long as it is not in the same compartment and it is well managed. Example: fire in a strip mall. The occupancy is well involved and the IC determines that the defensive strategy is indicated for the fire occupancy. The adjoining occupancies on either side of the fire are not on fire. The IC assigns companies to advance handlines into the exposures to keep the fire from extending. The IC assigns sector officers in both exposures and gives them the objective, "to keep the fire from extending through the attic and do search and rescue" (offensive).

Fire Command Instructor's Guide

Simulation Questions

The IC also has units assigned to knock down the main body of fire with master streams (defensive). Units operating on the main body of fire are also ordered not to operate their streams in the exposures because crews are operating inside of them.

Fire Command Instructor's Guide

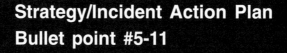

Simulation Questions

1. Describe how you plan on implementing/executing your strategy and plan for our fire.
2. What orders will you give to the companies/sector officers that you assign to the different key tactical positions?

■ Use the incident organization and communications to connect and act out strategy/plan.

The IC starts and stays in control when everyone operates within the confines of a well-managed system. The IC is the Grand Poo-Bah of the system. The radio is the tool the IC uses to manage incident operations. If the IC gets knocked off the air (for any reason), s/he no longer has the ability to manage the incident. The strategy, IAP and the subsequent assignments are shared and acted out when the IC verbalizes them (over the tactical channel) to his/her anxiously awaiting minions. This gets the show on the road.

The IC furthers his/her control of the operation when s/he decentralizes management of the hazard zone by assigning sector officer responsibilities. Sector officers operate in forward positions and control access in and out of the hazard zone for the points of entry in their assigned areas. They also generally have a better view of the conditions in their sectors than the IC does and are in a much better position to manage the safety of the people operating in the sector.

Fire Command Instructor's Guide

Strategy/Incident Action Plan
Bullet point #5-11

Simulation Questions

The IC provides the sector officer with the objectives for their sector. This becomes the starting point for conducting operations within the sector. As progress is made, objectives are met or conditions change (good or bad), the sector officer reports back to the IC. The IC processes the reports from all the operating sectors to manage both the strategy and the plan.

Fire Command Instructor's Guide

Simulation Questions

1. What sector assignments will you use to manage our fire?
2. What kind of command support does the initial IC (company officer) have?
3. How many resources can IC #1 effectively assign and manage?
4. What kind of command support does the IC #2 have?
5. How many resources can IC #2 effectively assign and manage?

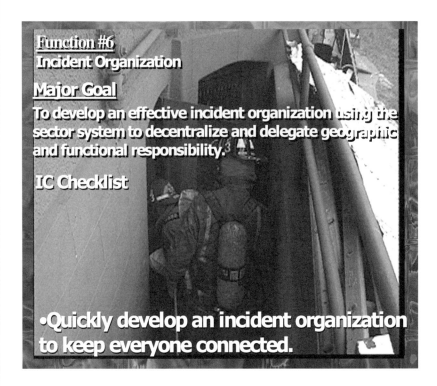

Function #6
Incident Organization

Major Goal
To develop an effective incident organization using the sector system to decentralize and delegate geographic and functional responsibility.

IC Checklist

•Quickly develop an incident organization to keep everyone connected.

■ Quickly develop an incident organization to keep everyone connected.

The IC does the first five functions of command to take control of the incident, identify the incident problem, figure out the best solution, communicate the problem and the plan to all the incident players, and assign resources to solve the problem. The IC accomplishes this within the first few minutes after arriving at the scene.

The IC should assign sector officer responsibility to the officers of the companies initially assigned to key tactical positions (the geographical places from where incident operations are launched). As the incident escalates, these positions should be upgraded to chief officers as quickly as possible.

Assigning sector officer responsibilities as part of the initial assignment to the officers assigned to these areas gets the IC's IAP implemented from the beginning of incident operations.

Fire Command Instructor's Guide

Simulation Questions

Company officers must manage and supervise their crews and assist, as needed, with carrying out task level activities. They are not in the greatest position to manage full-blown, large-scale sector operations, but they provide us with the best shot at getting initial sector operations up and running. When resource requirements or the hazard level in the sector escalates, the IC upgrades the sector officer position by transferring it to a person whose sole function is to be the sector officer (as opposed to a working company officer).

Fire Command Instructor's Guide

Simulation Questions

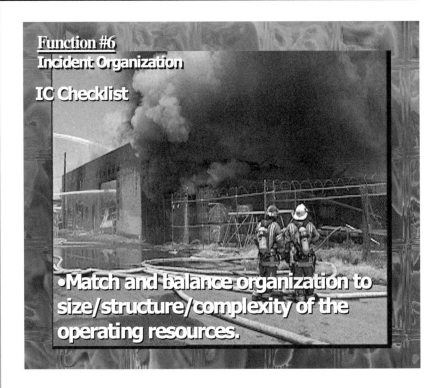

Function #6
Incident Organization

IC Checklist

•Match and balance organization to size/structure/complexity of the operating resources.

1. Based on the size and complexity of our fire, what sized management organization will be required (IC, IC staff, sector officers, etc.)--how many people?
2. How many of these managers will be physically involved in solving the incident problem (the tactical priorities)?
3. Does the IC have more people working than managing?

■ Match and balance the command organization to the size/structure/complexity of the operating resources.

Most offensive fire fights don't last very long. In many cases, the initial IC and sector officers (company officers) quickly solve the incident problem in their assigned sector areas, and the incident management system doesn't escalate beyond that level. This is a good thing.

When the incident does escalate beyond the control capabilities of the initial wave of problem solvers, it is time for the IC to begin reinforcing and expanding the command organization--on all levels. This needs to be a regular and ongoing part of the incident management system. If we use the same sized management structure to conduct hazard zone operations at a large and complex incident as we do for small, routine and "simple" events, we are asking for trouble.

The IC must always operate in a mode where s/he can manage and control the position and function of all assigned resources. The command organization must be able to at least match and pace the deployment of companies, personnel and other resources.

Fire Command Instructor's Guide

Function #6
Incident Organization

IC Checklist

•Forecast and establish geographic/functional sectors.

Simulation Questions

1. List the first four sectors that the IC will assign.
2. List all the remaining sectors that will be assigned--be sure to include functional sectors (i.e., rehab, etc.)
3. How many of these sectors, in question #2, are implemented or started automatically (per SOPs)?

■ Forecast and establish geographic/functional sectors.

Determining the key tactical positions is part of the IC's initial and ongoing size up. These key positions become the basis for assigning unit and sector officer responsibility. Forecasting these needs, as part of the initial size up, also gives the IC a general idea of how many resources will ultimately be needed. The IC prioritizes the order in which each of these areas is addressed.

Many of the functional sectors are automatically tied to alarm levels and resource requests. In many systems when a working fire is declared, as part of the initial report, resource and management response is automatically upgraded--a fire investigator, public information officer, utility truck (lights, SCBA air supply, etc.) and an extra engine company are dispatched.

In tactical positions, functional sectors are responsible for some type of support work. This ranges from water supply sectors searching out water sources, to vent sectors working on removing the products of combustion from the entire structure, or some part of the structure.

Fire Command Instructor's Guide

Simulation Questions

1. How fast was the IC able to get managers in place and operating in key tactical positions?
2. What is the span of control for the different sector officers?
3. Can sector officers assigned to non-hazard zone positions safely manage more people/ companies than hazard-zone sector officers--as an example can rehab manage more people than roof sector? Why can they?
4. Can the IC manage, maintain control, and support all of the sector officers required for this incident?

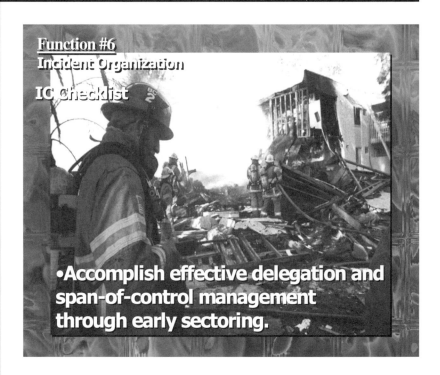

Function #6
Incident Organization

IC Checklist

•Accomplish effective delegation and span-of-control management through early sectoring.

■ Accomplish span of control management through early sectoring.

Early sectorizing provides the IC with two large benefits. First, it places management, supervision and a command partner in key tactical locations. This is the platform that the IC uses to expand his/her incident management capability.

The second benefit is that the tactical and task level responsibility is shifted from the IC to the sector officer for all the resources assigned to their sector. The IC will not remain strategic very long if s/he has to manage the operational details of a rapidly escalating event on a company-by-company basis. Assigning nine individual companies under the supervision of three different sector officers has the positive span of control effect to the third power (this is a snazzy way of saying it cuts the IC's span of control responsibility by two-thirds).

Fire Command Instructor's Guide

**Organization
Bullet point #6-5**

Function #6
Incident Organization
IC Checklist

•Correctly name sectors.

Simulation Questions

1. What did you name each sector?
2. Does the sector naming system make sense to all the incident participants--when you assign companies to a particular sector, will they have a general idea of the location of that sector?

■ Correctly name sectors and landmarks.

Correctly naming sectors eliminates confusion. When the IC assigns an engine and a ladder company to the north sector s/he shouldn't be surprised when they end up on the north side.

When the IC sectorizes, it has the effect of chopping the incident scene into more manageable pieces. The sectors become the work areas that companies are assigned to. This facilitates accountability for everyone. The IC eliminates any mystery when the sector name matches the geographic position of the sector. In some systems, where the community is laid out in rectangles, this is done geographically--north, south, etc. Other communities use letter designations--side A, side B, etc., as they go clockwise around the building.

At large scale incidents the IC may use landmarks to identify sectors (or other things, like cutoff points)--"Command to north sector... I want you to keep the fire from crossing Old Man Kelsey's bridge."

Fire Command Instructor's Guide

Simulation Questions

1. List the objectives given to each sector officer.
2. Do the objectives fit into your (the IC's) overall IAP?
3. How do you communicate this info to the sector officer (face to face, over the radio, etc.)?
4. Why is it important for sector officers to know one another's objectives?

Function #6
Incident Organization
IC Checklist

•Assign and brief sector officers

■ Assign and brief sector officers--provide objectives.

When the IC orders the initial company (and officer) to the incident scene, s/he provides the objective of that assignment. This is the way we begin incident operations. Very rarely, if ever, does the IC have a face-to-face briefing with the initial units assigned to the incident. As the operation continues and command officers are assigned to transfer sector officer responsibility from company officers, the IC will give a more detailed briefing. This may require a face-to-face meeting, especially if multiple sectors are being transferred.

Company officers who are initially assigned as sector officers should be given objectives that they can achieve. The IC must keep in mind that company officers will be involved in the physical, task-level activities required to achieve those objectives. They will also be operating in whatever conditions that are present in their area of operation. Fire companies focus on completing whatever task the IC assigns to them. If the IC orders an engine company, "Pull an attack line, complete search and rescue, and fire control on the third floor, you will be sector 3," that company will work towards those ends. If the third floor is 5,000 square feet and one quarter involved in fire, the company may work until they are almost out of air. The IC should always assign a task that the unit will have a realistic chance of completing.

Fire Command Instructor's Guide

Function #6
Incident Organization

IC Checklist

•Limit units assigned to sectors to five (5).

1. How many companies did you assign to each sector?
2. Can the sector officers effectively manage the number of people/companies that are assigned to them?
3. How does the IC improve the management capability within sectors?

■ Limit units assigned to sectors to five (5).

This number represents the upper end of the number units that a well-supported sector officer, who is operating outside the hazard zone, can manage in a forward position. A sector officer who is managing in this fashion is outside the products of combustion/hazard, has the ability to write down and track (from the outside) assigned units and to constantly evaluate the conditions and the effect of control efforts. In many systems later arriving command officers, who in many cases will have an aide assigned to them (built in command partner), fill these sector officer positions.

A company officer, who is managing and assisting his/her crew, inside the hazard zone has their hands full. It is absurd, dangerous, and incorrect to expect this company officer to effectively and safely manage five other units as an initial sector officer. The IC should limit the number of units assigned to sector officers that are operating in the hazard zone to between two to three (including their own crew). This decision is based on the sector officer's ability to see and stay in contact with assigned personnel, the level and potential of risk/hazard, and the ability to quickly leave (evacuate) to safe positions.

Some sectors can manage more than five units. These are typically sectors that operate outside the hazard zone. Staging and rehab are both examples of sectors that routinely will have more than five units assigned to them, particularly at large scale events.

Fire Command Instructor's Guide

Simulation Questions

1. Describe the process the IC uses to allocate resources to sectors.
2. How does the IC position uncommitted resources closer to the sectors (and the corresponding physical location) where they will be eventually needed?
3. Describe how these uncommitted resources can be utilized within the sector (i.e., RIC, tactical reserve, etc.).

Function #6
Incident Organization
IC Checklist

•Serve as a resource allocator to sectors

■ Serve as a resource allocator to sectors.

The IC must decide on the strategy, formulates an IAP, and make assignments (along with sector responsibility) to the key tactical positions. This front-end investment places sector officers in key operating positions. These sector officers are in the best position to carry out the IAP for their area of operations. The IC should avoid automatically assigning more resources to these officers/ positions. Once the system is implemented, sector officers should report back to the IC on the conditions, the actions they are taking, and the needs in their sector. This allows the IC to operate on the strategic level, serving as a resource allocator to the sectors. The IC talks (gives orders and makes assignments) to get the organization in place and operating. After the organization is in place, the IC should shift from primarily talking to mostly listening, and reacting to reports. In many cases, this is all that is required to keep the strategy and IAP current.

Fire Command Instructor's Guide

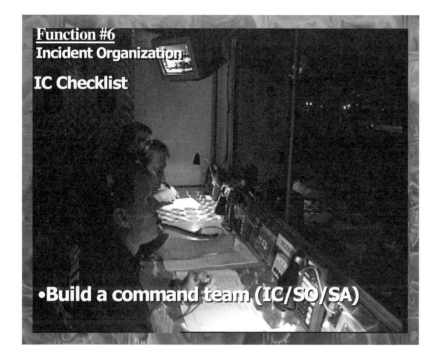

Function #6
Incident Organization
IC Checklist

•Build a command team (IC/SO/SA)

Simulation Questions

1. Describe how your system supports the IC.
2. Describe the evolution of that support-- who provides it, how long it takes, the command post command will operate out of, etc.
3. Can you assemble a command team, in a timely fashion, for our fire?

■ Build a command team (IC/SO/SA).

Command Teams are a regular piece of the management/ support system escalation that we use in our department. This is the way we implement them:
- initial arriver (generally a company officer) assumes command
- first arriving chief transfers command, becomes the IC
- first arriving deputy chief assumes support officer
- second arriving deputy chief becomes senior advisor

The command team operates out of the command van.

This is one way, that a fire department builds a local command team. Since the organizational details will vary from one department to the next (number of responders, different rank structure, arrival order, etc.) we must localize the way our departments manage "command" and support the IC.

Fire Command Instructor's Guide

Simulation Questions

The support officer does just that--supports the IC. S/he does this by filling out and managing the tactical worksheet, verifying that the strategy and IAP match the conditions, and keeping everyone else away from the IC (shielding the IC from distractions).

The senior advisor manages the sections. Logistics and safety sections are normally the first two section positions that we implement. The senior advisor serves as the connection between the IC/support officer and the sections. The senior advisor is also responsible for managing the command post, connecting the incident to the rest of the city (operationally and politically), verifying that the strategy and IAP match the conditions and expanding the organization, as needed.

Fire Command Instructor's Guide

Organization
Bullet point #6-10

Function #6
Incident Organization

IC Checklist

•Build outside agency liasion and public information into the organization.

Simulation Questions

1. What other agencies will be required for our fire?
2. How does the IC plug these outside agencies into his/her IAP?
3. Who will serve as the incident PIO?
4. What should we tell the community about this incident?

■ Build outside agency liaison and public information into the organization.

For this to be effective, the system must be in place (and practiced) ahead of the event. We routinely conduct incident operations with a variety of other agencies. The activities of all these different organizations must be coordinated and integrated. If the fire department is taking the lead in incident operations, the IC will be in overall charge of all incident activities. This includes where other agencies are and what they are doing. If a different agency has taken the lead role as the overall incident IC, the fire department reports to them. Instances of these types of incidents include:

• The police having control of a scene that is not secure-- people are running around with guns or other weapons and it is not safe for firefighters to operate at the scene.
• The utility company is in charge of incident operations--a major gas leak, electrical problem, etc.
• After the incident care and support. We have stabilized the incident problem and are ready to turn after incident
• Customer support over to another group/agency (i.e., Red Cross, Chaplain, Social Services, etc.)

Organization
Bullet point #6-10

Simulation Questions

The fire department IC is always responsible for managing the fire department personnel and resources. When another agency is taking the management lead, the fire department IC remains in control of FD resources. The IC (or his/her representative) liaisons with the lead agency IC or liaison.

We routinely operate at incidents where we must coordinate the activities of several different response agencies. If the incident has several different agencies that play a large role in incident operations, we must coordinate the activities between the different agencies. One of the most effective ways to accomplish this is through the use of liaisons from each agency in the command post.

The last person who needs to be doing news interviews during the active phase of incident operations is the IC. Having a full-time PIO provides the news crews a familiar person to provide them with the information they want.

Organization
Bullet point #6-11

Function #6
Incident Organization

IC Checklist

•Operate on the strategic level -
support tactics - task levels.

1. How quickly will you have a strategic level IC in place (an IC that will operate in the command mode)?
2. Does the IC manage in a way that supports the units that are completing the tactical priorities?
3. How does the IC improve the safety of the hazard-zone workers?

■ Operate on the strategic level--support tactics--task levels.

The IC builds the appropriate sized incident organization so s/he can operate on the strategic level. The strategic level is responsible for doing the functions of command. The important concept in the three organizational levels (strategic, tactical, and task) is that the strategic and tactical levels exist to support the goals of the task level. The task level is where the customer gets saved, their stuff is protected, and the problem is solved. The task level activities will always be more effective and safer when it is managed by the right amount of strategic and tactical level support.

The IC identifies the overall strategy and develops and implements an IAP based on that strategy. The IC assigns tactical level responsibilities to tactical level sector officers, who supervise and manage where the work is being performed. The task level is the actual application of the work.

Simulation Questions

1. Once sectors are established and sector officers are briefed does the IC allow the sector officer to manage their sector?
2. How do sectors communicate their activities with one another?
3. How does the IC coordinate sector activities?
4. Which sectors need to have their tactical activities coordinated at our fire?

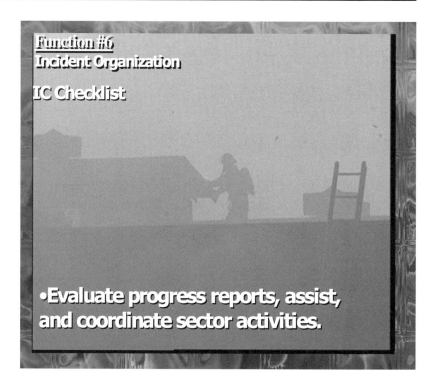

Function #6
Incident Organization

IC Checklist

•Evaluate progress reports, assist, and coordinate sector activities.

■ Evaluate progress reports, assist, and coordinate sector activities.

The IC assigns sector responsibilities and objectives to put the IAP in motion. After the IC gets the right amount of resources and management in place, s/he is in the strongest position to manage the continuation of the attack. After incident operations have begun, the IC uses progress reports to keep the strategy and IAP current. The IC also coordinates activities between sectors.

Fire Command Instructor's Guide

Function #6
Incident Organization

IC Checklist

•Implement management sections (ops - planning - logistics – administration - safety) and branches to provide support and connect operational and command escalation - delegate.

Simulation Questions

1. What, if any, section positions will the command team implement?
2. Will branch officers be required for our fire?

■ Implement management sections (ops, planning, logistics, administration, safety) and branches to provide support and connect operational and command escalation--delegate.

The IC and command team will remain effective if they are able to operate on the strategic level. If the command team allows themselves to get bogged down in the details of the tactical and task level, incident operations as a whole will suffer. The command team must use the different parts and pieces of incident organization to escalate operations and delegate the detail management.

The command team's main goal and focus must be the management of the workers who are operating in the hazard zone. This includes providing whatever technical support (i.e., special operations, hazmat, technical rescue, etc.) that is necessary.

Large scale and complex incident operations will require a larger command staff to manage the added organizational positions that must be filled. These positions provide logistical support, planning and administrative functions, safety officers and the branch officer positions, where needed.

Simulation Questions

1. How many communications partners will the IC have for our fire?
2. Describe the streamlining effect that implementing a proper sized incident organization has on radio communications.
3. Which sectors have a critical safety need for instant communications with the IC?

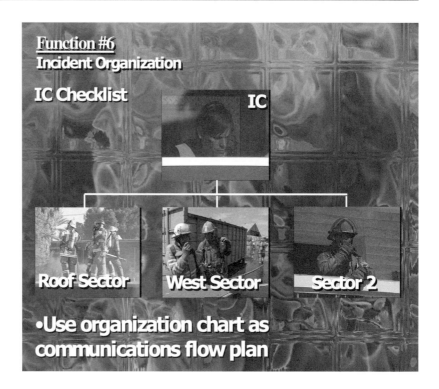

Function #6
Incident Organization

IC Checklist

IC

Roof Sector   West Sector   Sector 2

•Use organization chart as communications flow plan

■ Use the organizational chart as the communications flow plan.

The communications flow chart should overlay and reflect the organizational chart. The IC is will generally communicate with a couple of engine companies and a ladder company at the scene of a small and "routine" incident over a single tactical radio channel. The incident organization should reflect this. One would not expect to find a full command team with all the section positions implemented at the scene of a room and contents fire in an ordinary house. On the other hand, it is inappropriate for a single IC to try and manage a large or complex incident using the same communications and organizational model that s/he would use to manage a house fire.

Fire Command Instructor's Guide

Organization
Bullet point #6-15

Function #6
Incident Organization

IC Checklist

•Allow yourself to be supported in the process.

Simulation Questions

1. Does the IC take direction from the command team?
2. How are strategy and IAP conflicts resolved between members of the command team?

■ Allow yourself to be supported in the process.

The IC sends fire companies to support other fire companies. It only makes sense that we send more chiefs to support the IC. This is no way reflects on the capability of the IC. We do this to make the IC more effective. Allowing yourself to be supported requires a special level of ego control. As we use and operate within a system that makes IC support a regular and automatic part of a regular local incident management system, this team approach becomes less of an issue.

When the support officer shows up and asks the IC, "What's your plan?" It's not meant to be a personal affront to the IC's skill level. It's a question that should cause both the IC and the support officer to clinically analyze the critical factors, verify that we are operating in the correct strategy, that the actions match the conditions, and other critical things that go into the incident management process.

Simulation Questions

1. Describe the review, evaluation and revision system the IC uses to make changes and adjustments to the IAP.

2. Does this system allow the IC to quickly change the position and function of operating resources?

Function #7
Review - Revision
Major Goal
To complete the steps required to keep the strategy and action plan current.
IC Checklist
•Regular command system elements established in the beginning provide the framework for mid-point review - revision.
*strong standard command
*sectors
*SOPs
*risk management plan
*strong communications
*standard strategy/action planning

■ Regular command system elements established in the beginning provide the framework for midpoint review--revision:

- strong standard command
- sectors
- SOPs
- risk-management plan
- strong communications
- standard strategy/action planning

The ongoing review process is nothing more than a continuation of the same system that we used to get the operation up and running. If the IC didn't do the first six functions of command during the initial stages of incident operations, it makes it next to impossible to make necessary changes and adjustments later on in the incident.

Fire Command Instructor's Guide

Review, Evaluation, Revision
Bullet point #7-2

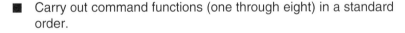

•Carry out command functions (1-8) in a standard order.

1. List the problems that can arise when the IC skips doing one of the functions of command?

2. Is it possible to make operational changes (strategic and with the IAP) when the IC didn't do all of the functions?

■ Carry out command functions (one through eight) in a standard order.

The initial IC does the first five functions of command during the first few minutes of initial incident operations. The functions are laid out and presented in the natural order that the IC does them. This provides a standard system that we use to manage the emergency service that we deliver. The regular and ongoing use of this standard system builds familiarity among all the incident players. This causes our incident operations to become more effective and safer. The command system is designed to achieve the standard tactical priorities in a regular fashion. The IC uses the system to take control of the incident, determine the scope and size of the incident problem, take control of incident communications, request for and assign resources, determine the appropriate strategy and corresponding IAP, decentralize the management process by assigning sector officer responsibilities, provide for the constant review and evaluation of the incident operations, and make any required changes, to transfer command to later arriving officers, and to terminate command when the event is completed.

Function #7
Review - Revision

IC Checklist

•Receive, confirm, and evaluate
conditions - progress reports.

Simulation Questions

1. What information does the IC want the progress report to contain?
2. List a probable progress report from the adjoining occupancies (the exposures).
3. List a probable progress report from the fire occupancy.
4. How do progress reports effect the strategy, IAP and firefighter safety?

■ Receive, confirm, and evaluate conditions--progress reports.

The IC uses the visual size up of conditions and progress reports from companies and sectors as the basis for command function #7. The visual information is limited to the command position of the IC (physical location). If the IC is operating in the fast-attack mode (company officer) his/ her view will be limited to whatever area they are currently working in. If the IC is operating in the command mode, they should have a pretty good overall view of two sides of the incident scene.

The goal of the system is to place an IC in the command mode as quickly as possible. The IC can then monitor (look, listen, process, react) the overall effect that operations are having on the incident problem. The IC gets IAP info back from the operating companies/sectors. This information should include a description of the critical factors and start to "uncover" any critical unknowns in that sector. Example: the IC is commanding a structure fire.

Fire Command Instructor's Guide

Simulation Questions

Part of his/her IAP is to verify that the fire has not extended to the attic. The IC has made this one of the top priorities for a continued offensive fire attack and assigned this to both the roof sector and north sector--the two most likely places to find out this information in a timely fashion. The IC can see the overall conditions from the command post and can tell if conditions are getting better or worse. Depending on the building design and attic configuration, conditions may be getting worse in the attic space, but they will not show themselves (at least from where the IC is sitting) until it is too late to do anything about it (other than running for your life if you are anywhere near the gravitational field that the roof effects). The roof and north sector officers must quickly determine this attic information and then get this information back to the IC so s/he can keep the IAP current, basing actions on the most current conditions.

Fire Command Instructor's Guide

Simulation Questions

1. How often does the IC (and everyone else) evaluate the effectiveness of the IAP and the effect that our actions have on conditions?
2. Does the strategy and IAP get re-evaluated after one of the tactical priorities have been completed?
3. We have put an IAP in place to control the fire in our example, have assigned the initial attack wave, and are five to eight minutes into the operation. Conditions are getting worse. How does the IC react and deal with this?
4. Is the IC in a position to quickly switch to plan B?

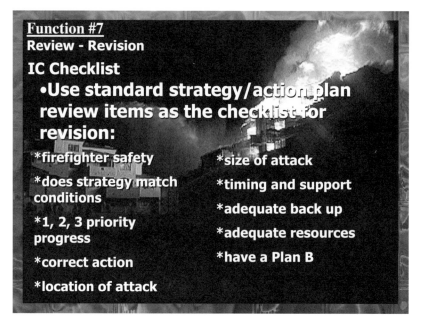

Function #7
Review - Revision

IC Checklist
•Use standard strategy/action plan review items as the checklist for revision:

*firefighter safety

*does strategy match conditions

*1, 2, 3 priority progress

*correct action

*location of attack

*size of attack

*timing and support

*adequate back up

*adequate resources

*have a Plan B

■ Use standard strategy/action plan review as the checklist for revision:

- firefighter safety
- does strategy match conditions
- first, second, third priority progress
- correct action
- location of attack
- size of attack
- timing and support
- adequate back up
- adequate resources
- have a plan B

IAP #1 is the initial radio report. It identifies that someone has arrived to the scene, describes the incident problem and the action taken, the incident strategy, and identifies who the IC is. The minute or so that the IC spends on making a good initial report wraps the first five functions of command together and serves as the foundation for the first IAP. In many (most) cases, this solves the local incident problem. In these cases, the IAP makes a natural transition from a search and rescue/fire control

Fire Command Instructor's Guide

Simulation Questions

(once these objectives are achieved) focus to a checking for fire extension, smoke removal, controlling the loss, and customer stabilization. After these operational targets are achieved, the IAP shifts again, focusing on rehabbing personnel, fire investigation, securing the property, and turning it back to the customer and making sure the customer has the necessary after the fire support. The IC begins the operation with an IAP that makes two or three operational shifts for a simple contents fire, requiring the commitment of three or four tactical units.

When the initial IAP doesn't solve the incident problem, the IC must revise it based on all the bullet points listed above. This process should start with making sure that firefighters are operating in safe positions and that the strategy matches the conditions. For offensive situations, the next item to evaluate is the progress towards completion of the primary search. The IC has to figure out if it is more effective/safer to eliminate the problem then complete the search, or to finish the search prior to shifting the operational focus to fire control. This decision is based on the critical incident factors (building size, age, layout, occupancy type--fire location, extent, intensity, fuel type and load --available resources, etc.).

This is a very dynamic process and requires a fully conscious IC.

Fire Command Instructor's Guide

Simulation Questions

1. When formulating the safety profile of our example, list the set of factors that would lead the IC to choose the offensive strategy.

2. What set of factors would cause the IC to choose the defensive strategy?

3. A lot of people talk about having a back-up plan, or plan B (we mention it several times ourselves). Explain how the IC is able to quickly move from plan A to plan B in such a manner that everyone's safety is verified, and there is not a large gap in operational effectiveness (we don't waste a lot of time going from one plan to the next while the building burns).

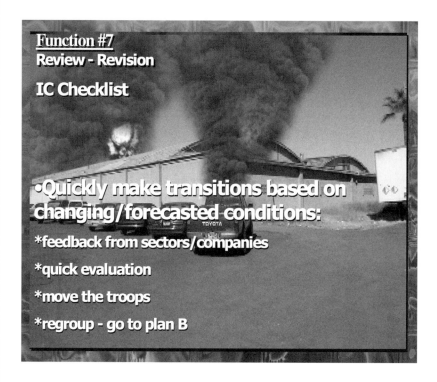

Function #7
Review - Revision

IC Checklist

•Quickly make transitions based on changing/forecasted conditions:

*feedback from sectors/companies

*quick evaluation

*move the troops

*regroup - go to plan B

■ Quickly make transitions based on the safety profile of changing/ forecasted conditions:

- feedback from sectors/companies
- quick evaluation
- move the troops
- regroup--go to plan B

After the resources that the IC assigned to the key tactical positions have had a chance to get into place and begin operations, the IC must quickly determine the effectiveness of their actions. A well-executed offensive fire attack will generally quickly control the fire. If the problem (the fire) continues to grow despite the offensive control efforts, it is a sure sign that the current plan is not working. The IC must determine if the problem can be solved by reinforcing current positions, if a key attack position has not been addressed, or if the fire is too big to control from interior positions with hand lines. This decision must factor in

Fire Command Instructor's Guide

Simulation Questions

how long it will take to get resources to the needed positions, and how long it will take to evacuate, and account for interior crews if conditions continue to worsen.

If the IC determines that it is no longer safe for fully protected firefighters, who are operating within a well-managed system to operate in the offensive strategy, it is probably a safe bet that there is nothing left to save. When there is nothing left to save, the risk-management plan tells us we will not take a risk.

Simulation Questions

1. What does IC #2 want in place when s/he transfers command from IC #1?
2. List the major problems that we experience if IC #1 does not effectively do the command functions.
3. How long does it take IC #2 to recover and stabilize incident operations when s/he inherits a poorly managed front end?
4. How does IC #2 rein things back under control?
5. If IC #2 has three engines and a ladder assigned to the scene of our fire and command is out of whack, is it possible to get these resources back into the incident management fold while continuing to operate in offensive positions with the conditions shown?

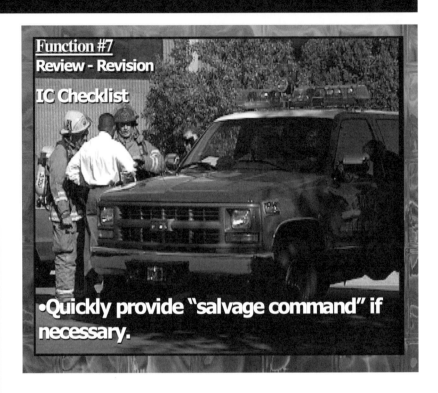

Function #7
Review - Revision
IC Checklist

•Quickly provide "salvage command" if necessary.

■ Quickly provide "salvage command" if necessary.

One of the key size-up items that must be considered by IC #2 (normally the "chief") is if the initial attack wave is well managed and under control. This later arriving IC must look at the incident conditions and the actions that have been taken to control them through a fresh set of eyes. The new IC should be able to figure out in short order whether or not things are going according to SOPs, any kind of sane and lucid IAP, and to the new IC's expectations.

The second arriving IC should be the most skilled and effective command person on the scene (at that point in the operation). They needed to have the ability, confidence, presence, etc. to quickly take command (transfer) and do whatever is necessary to get the incident back under control, if things are out of balance (screwed up). The new IC (along with all other humans) may not be able to have a positive impact on the incident conditions. The one area they can impact, and the biggest reason we send them to the scene, is to manage the strategy. If IC #2 pulls up and finds the initial attack wave in a earnest offensive struggle with what is clearly a defensive, no win, fire, s/he must quickly pull the plug and get everyone to safe defensive positions. This is the most important IC activity--making sure everyone finishes their shift.

Fire Command Instructor's Guide

Continue, Transfer, and Terminate
Bullet point #8-1

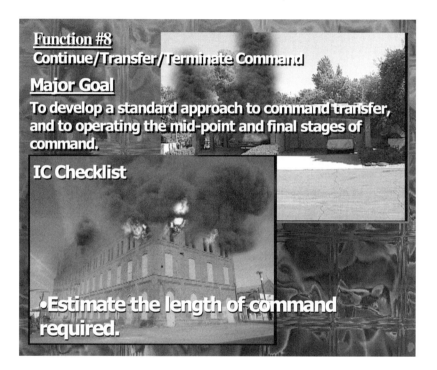

Function #8
Continue/Transfer/Terminate Command

Major Goal
To develop a standard approach to command transfer, and to operating the mid-point and final stages of command.

IC Checklist

•Estimate the length of command required.

Simulation Questions

1. How long will IC #1 be in command?
2. How many units can IC #1 effectively manage?
3. How long will IC #2 be in command?
4. How many units (or alarms) can IC #2 effectively manage (out of his/her response vehicle)?
5. At what point is IC #2 supported?
   a. Does IC #2 have a FIT (i.e., aide, tech or whatever else they may be called)?
   b. Describe the effect having a FIT has on the IC's capability to command.
   c. Describe the next level of support or escalation of the system and the time frame that it occurs in.

■ Estimate the length of command required.

This helps to establish and identify several incident management issues, estimating the size of command organization that will be required and determining how long the incident will last.

This size-up factor is primarily based on the size and scope of the incident. The initial IC is normally a company officer and is busy with the first five functions of command and assisting his/her crew with task-level activities. When the initial IC determines that more resources along with the corresponding higher amount and level of command, it comes out as a request for more resources (striking additional alarms). This notification is generally made as part of the initial radio report.

The IC should forecast how long incident operations will last and how big a command organization will be required. This size up should occur quickly, at least by the time command has been established in a command post. If the incident is going to last

Simulation Questions

beyond a certain length of time (longer than a single command team can reasonably manage), a schedule should be developed that provides for the rotation of the command team and any other members who are staffing positions that will be used throughout the event. The length of time that an IC and the rest of the command team remains in command of an event is based on time and intensity. Long duration, slow-moving events (burning debris piles with no exposures, defensive fires with no exposure problems, etc.) are not as stressful as incidents that are more complex, with personnel operating in a hazard zone.

Fire Command Instructor's Guide

Continue, Transfer, and Terminate
Bullet point #8-2

Function #8
Continue/Transfer/Terminate Command

IC Checklist

•Consider the time for completing each tactical priority.

■ Consider the time for completing each tactical priority.

Each of the tactical priorities represents the core for the IAP at any given point of incident operations. The IC bases deployment (assigning units) to the incident needs. One of the constant deployment "companions" is the IMS component required to manage all of the assigned resources. In the preceding bullet point the IC begins incident operations estimating the total length of time required to complete incident operations. This bullet point starts to break the entire incident operation into the smaller pieces (and time frames) that correspond with the continually evolving IAP that the IC develops, implements, manages, and revises over the course of the incident.

In many cases the IC will not begin incident operations with all the resources that will ultimately be required. Estimating how long each tactical priority will take, along with how many people/ crews they will require, gives the IC a general idea of how many

Simulation Questions

1. How long will it take to get an "all clear" in all required areas?
   a. How many units will be required?
   b. Who will manage and supervise the completion of this tactical priority?
2. How long will it take to get an "under control" in all re- quired areas?
   a. How many units will be required?
   b. Who will manage and supervise the completion of this tactical priority?
3. How long will it take to get a "loss stopped" in all required areas?
   a. How many units will be required?
   b. Who will manage and supervise the completion of this tactical priority?
4. How will the IC manage the after fire support (customer stabilization) that the property/business owners will require?

101

Continue, Transfer, and Terminate
Bullet point #8-2

command players incident operations will require. If the IC estimates that the bulk of the work will be done in ten minutes, by three or four crews, the response time of the command responders (chiefs) needs to be factored in. If it takes ten minutes for the next-due chief to get to the scene, company officers will be filling those roles (as sector officers). As long as the resource levels that they are responsible for don't go beyond their management capability, this is okay. If the incident is going to escalate beyond this time and resource estimate, the IC must forecast the need and get the required command resources (chiefs) coming. Getting the required resources to the scene ten minutes after you needed them produces nothing but frustration.

Fire Command Instructor's Guide

Continue, Transfer, and Terminate
Bullet point #8-3

Function #8
Continue/Transfer/Terminate Command

IC Checklist

•Consider life safety, fire area profile and incident/fire conditions.

Simulation Questions

1. What is the life safety size up for our fire?
2. What is the incident/fire area profile for our fire?
3. What is the incident/fire conditions size up for our fire?

■ Consider life safety, incident/fire area profile, incident/fire conditions.

Life safety is the first and foremost tactical priority. Incidents that involve large and extended life safety operations are big deals that require lots of command support. These types of events often times involve medical treatment and transportation components, which must be managed and incorporated into the IC's IAP. Every time another major management area is added, it takes more of the IC's limited attention (we're all human) and time. Before incident activities overwhelm the IC, s/he must be supported because when the IC vapor locks, incident management goes out the window.

The incident/fire area profile represents the possible size, scope and potential of the incident. This plays a large role in determining the corresponding resources that will be required to bring the event under control. The layout of the incident terrain will determine action taken, the length of time it will take to engage the incident problem (how long it takes to get in and how long it takes to get out), the resource level that will be required to safely and effectively solve the incident problem and the corresponding

Continue, Transfer, and Terminate
Bullet point #8-3

amount of command support players that are needed to actually manage incident operations.

The incident/fire conditions describe the incident problem. When we marry the problem to the incident profile, it should lead us to the required action. These two items represent the physical environment where we deliver service and conduct operations. A mathematical formula for incident operations looks like this: Incident conditions + incident profile + tactical priorities + risk management plan = action taken. Action taken (this includes how much and what kind of action is needed and where that action must take place) = resource level. The IC must always be able to control the position and function of all assigned resources. S/he must build a command organization that corresponds to this incident management constant.

Fire Command Instructor's Guide

Continue, Transfer, and Terminate
Bullet point #8-4

Function #8
Continue/Transfer/Terminate Command

IC Checklist

•Develop an organization that outlast the major event.

Simulation Questions

■ Develop and support an organization that outlasts the event.

It doesn't make any sense to apply a fifteen-minute operation to a thirty-minute problem. In fact this is an extremely dangerous way to conduct day-to-day operations. It is also something that most of us do. This is because about 98% of the time our fifteen-minute operation solves the problem. The possibly lethal payback comes from the 2% of the incidents where our standard, routine, and well-executed fifteen-minute plan doesn't work. The IC's standard incident solving approach must be supported with a steady, "in time" stream of resources. Our ten-to-fifteen-minute plan is usually pulled off with two to four engine companies and one or two ladder companies. This is adequate if one or two engines and a ladder company are all it takes to control the incident problem. The problem gets ugly when they don't, and the IC must then assign the remaining tactical reserve to reinforce existing positions or fill in any uncovered positions. This depletes the tactical reserve (staged units) and puts the IC (and everyone else) in a position where they have to play catch up. Now the IC has to request more resources and wait while these new

Simulation Questions

1. How many re-
    sources (alarms) will
    the IC need to
    effectively deal with
    the incident prob-
    lem?
2. How much com-
    mand support/
    incident organiza-
    tion will the IC need
    to effectively man-
    age the required
    resource level?

Continue, Transfer, and Terminate
Bullet point #8-4

Simulation Questions

units are dispatched and respond to incident staging areas, which will always take more time than the IC can afford.

The other problem associated with this approach is the way these two different kinds of incidents must be managed. If the IC isn't doing the required set of management activities that are required to escalate operations, it becomes almost impossible to jump from a simple two-engine, one-ladder operation into a bigger deal, more dangerous six-engine, three-ladder fire fight. The IC must begin and continue operations in a way that always allows for the continued escalation of operations.

Fire Command Instructor's Guide

Continue, Transfer, and Terminate
Bullet point #8-5

Function #8
Continue/Transfer/Terminate Command

IC Checklist

•**Assume, maintain and upgrade effective command positioning.**

■ Assume, maintain, and upgrade effective command positioning.

Offensive incident operations will usually begin with an IC that operates in the fast-attack mode (company officer). This ends when the incident problem is solved or command is transferred to an IC who will operate in the command mode (usually a chief operating in a command vehicle). Command is reinforced as later arriving chiefs (or whatever rank your system uses) arrive on the scene and support the IC. Depending on arrival order, rank and SOPs, these later-arriving officers can become the support officer and/or senior advisor, establishing a command team. The command team will normally be operating out of a larger "command van" command post. As later arriving officers become available, they can be used to fill sector officer, section chief or whatever position is needed at that point in the operation.

The system must be designed to support the IC in managing any type and size of incident that your department responds to. The system must progressively and naturally expand to improve the position of the IC and the level of IC support.

Simulation Questions

1. Describe the transfer of command process that your department will use for our fire?
2. How many times will you transfer command?
3. How many different individuals will serve as the IC?
4. How does the system identify the current IC?

Fire Command Instructor's Guide

Simulation Questions

Continue, Transfer, and Terminate
Bullet point #8-6

1. Describe your communications plan for this incident.
2. How many radio channels will you utilize for this incident?
3. How many radio channels can one person effectively manage?
4. How many radio channels will the IC be directly responsible for?

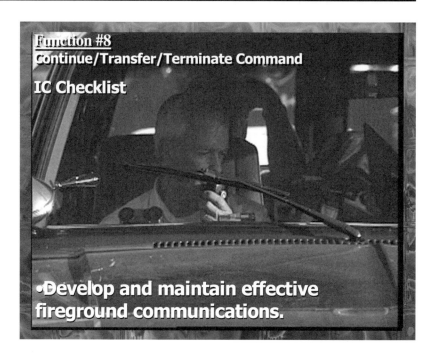

Function #8
Continue/Transfer/Terminate Command

IC Checklist

•Develop and maintain effective fireground communications.

■ Develop and maintain effective fireground communications.

Our commo capability is closely connected to command positioning. The fast-attacking IC will be running the incident over a portable radio. As command is transferred to an IC that will operate in the command mode both the radio (more powerful, headset, etc.) and the communication position (inside a vehicle) of the IC improves. As a command team is assembled, the command post is upgraded (in many systems) and there are more people directly supporting the IC and incident communications. This may include using multiple channels to run tactical operations, logistics, safety, etc.

The IC must be supported by a system that allows him/her to stay in constant, undistracted contact with all the companies that are operating in the hazard zone.

Fire Command Instructor's Guide

Function #8
Continue/Transfer/Terminate Command

IC Checklist

•Keep the incident action plan going (and growing if necessary).

Simulation Questions

1. What was IC #1's chosen strategy and IAP?
2. What is IC #2's strategy and IAP?
3. Which one is more focused on completing the tactical priorities for the entire incident scene?
4. Which one is more focused on tactically solving the problem in the immediate hazard area?
5. Which one is easier to expand and revise?

■ Keep the incident action plan going (and growing, if necessary).

This is the reason we make huge investments in and send ICs to incidents. The IC uses the first five functions of command to get the operation in place. He/she uses the sixth function to subdivide the incident scene and assign responsibility to manage the different areas and functions that must take place to achieve the incident goals (tactical priorities). The seventh function is where we evaluate our effectiveness and make any necessary changes. This function deals with how we support the IC, keeping him/her in control of incident operations.

A lone IC is not able to manage a rapidly escalating event that has three different operations going on (a fire, evacuation, and mass casualty, as an example) in twelve different sectors, across a single incident site. A general rule of thumb: the amount of IC support is directly relational to the size, hazard, and complexity of the IAP, and the number of required resources.

Fire Command Instructor's Guide

Simulation Questions

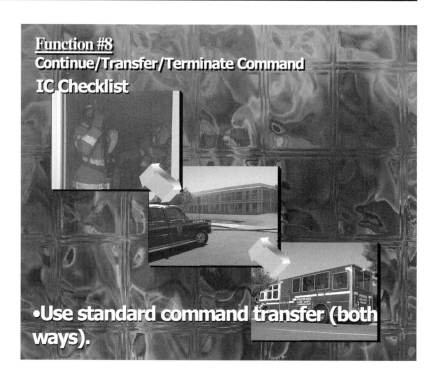

Function #8
Continue/Transfer/Terminate Command
IC Checklist

•Use standard command transfer (both ways).

1. Did you use a standard system to transfer command up the chain of command?
2. Describe the system you will use to de-escalate command.

■ Use standard command transfer (both ways).

We transfer command (from a company officer to a chief officer) to match the level of command required to conduct safe and effective incident operations. This program has gone to great lengths to detail and explain the benefits of upgrading, reinforcing, and supporting command (the IC). We must use a regular system to do this. That system must be based on the profile of our local resources and operating characteristics. It cannot be a mystery about who the IC is or how command gets transferred from one officer to the next. When command is transferred from the initial IC (often operating in the fast-attack mode) to the second-arriving IC (usually the chief), it needs to be clear that command has been transferred from one person to the next. The transfer system must cause one person to stop being the IC and for the next person to become the IC.

This transfer process works best when the outgoing IC and incoming IC have a face-to-face transfer meeting. This is not possible when IC #1 is operating inside a burning building and IC #2 is locked inside a command vehicle, two-hundred feet away.

Fire Command Instructor's Guide

Simulation Questions

This transfer will take place over the tactical channel. The new IC needs to contact the current (soon to be former) IC and let him/her know that s/he is on the scene and will transfer command. In many cases the transferring (new) IC will assign the old IC sector officer responsibilities for his/her area of operation.

After command is transferred, it is a good practice to contact the dispatch center and give an update and announce that you have taken command. Example: "Battalion 3 to dispatch...be advised that we have a working fire in the rear of a small commercial building. We have an all clear and crews are making good progress towards fire control. We are operating in the strategy and Battalion 3 will be command."

As incident operations wind down, command is transferred from the current IC to a company officer (or other person) who will be the last one to leave the scene. This is normally done at the very end of the incident, after all the tactical priorities have been achieved, just prior to terminating the active incident. We use the same system to de-escalate command as we did to escalate it. The only difference is we are reversing the rank and position of the IC (in many cases) --matching an on-scene IC to the level of command that is currently required.

Continue, Transfer, and Terminate
Bullet point #8-9

Simulation Questions

Function #8
Continue/Transfer/Terminate Command

IC Checklist

•Provide rehab, rotation, and relief for the IC and command staff.

1. Do you think that the IC/command team will require rehab, rotation and relief for this incident? Explain.
2. Describe the work cycle your department uses for the IC and command staff.

■ Provide rehab, rotation and relief for the IC and command staff.

This should be a regular part of extended operations. The IC, command team, sector officers, section chiefs and everyone else who is operating at the incident scene will need periodic rehab, rotation, and relief during the course of the incident.

Extended operations (for fire incidents) are generally slower moving, defensive events. During these types of operations (no one operating in a hazard zone), a well-supported IC may be able to stay in command for several hours or more at a stretch with an occasional break for stretching and other bodily functions and needs. Incidents that will take place over many hours (or days) will require operational and command periods along with some type of rotational roster. Unusually stubborn incidents, where hours turn into days, require cycling companies (and shifts) in and out of incident operations. Traditionally the first wave will work for an appropriate period of time (working in operational positions, cycling through rehab and back into operational positions) then they will be relieved and put back into service. It makes sense to replace the entire initial wave at the same time. This includes rotating the current IC and replacing him/her with a new

Fire Command Instructor's Guide

Continue, Transfer, and Terminate
Bullet point #8-9

one. This will be a big part of the IC's IAP--coordinating the relief for the initial assignment. If the incident will last for several days, some type of regular rotation schedule should be formulated to provide relief at regular intervals (send fresh replacements every four hours or whatever makes sense). This allows the dispatch center to schedule move ups and coordinate service delivery more effectively for the rest of the community. It also provides assigned crews a time frame for how long they will be operating at the incident scene.

Fast-moving incidents with a high risk to life safety are stressful and can wear out the IC and command team very quickly. While these types of incidents can be emotionally draining, they tend to be over fairly quickly. The best approach for managing these types of events is to support the IC. An IC working in a well-managed and staffed command post is in the best position to manage a fast-moving incident with severe consequences. When this type of incident turns into an extended operation, the IC should be rotated out and rehabbed as soon as it is feasible to do so. This is much easier to do when the IC is operating as part of a command team because two other people are familiar with the operation.